DATE DUE

Nov 13 '72			

TECHNOLOGY GAP IN PERSPECTIVE

TECHNOLOGY GAP IN PERSPECTIVE

Strategy of International Technology Transfer

By

Daniel Lloyd Spencer

SPARTAN BOOKS

New York • Washington

Library of Congress Catalog Card Number 70–118987
International Standard Book Number 0–87671–400–9
Printed in the United States of America.
Sole distributor in Great Britain, the British Common-
wealth, and the Continent of Europe.

Macmillan & Co. Ltd.
4 Little Essex Street
London, W. C. 2

February 8, 1967, © Punch, London.

A BRITISH VIEW: In February, the magazine Punch published this cartoon, an allusion to the migrations of scientists and technologists from Asia to Britain, and from Britain to the United States. But concern over the 'technological gap' goes beyond such shifts, and takes account of the U. S. as a potential 'technological empire.'

. . . I accost an American sailor, and I inquire why the ships of his country are built so as to last but for a short time; he answers without hesitation that the art of navigation is every day making such rapid progress that the finest vessel would become almost useless if it lasted beyond a certain number of years. In these words, which fall accidentally and on a particular subject from a man of rude attainments, I recognize the general and systematic idea upon which a great people directs all its concerns.

—Alexis De Tocqueville, 1840

Alexis De Tocqueville (trans. by Henry Reeve), *Democracy in America*, 2 vols. (New York: D. Appleton and Company, 1901), I, 516.

CONTENTS

PART II
POLARITIES OF THE INTERNATIONAL TECHNOLOGICAL SYSTEM

PART III
CHANNELS AND CONSEQUENCES OF WORLDWIDE TRANSFER

PREFACE

Some time ago, the author was called on to provide a "crash" comment on a policy matter involving the "technology gap" and other related problems discussed in this study. Having been working in a specialized area involving technology transfer through the military channel, I found the sudden projection of specialized research on a larger canvas very challenging. It quickly became apparent that the subject of international technological gaps and flows was much larger than anything which could be attempted in a summary paper. Although the specific task was performed in some fashion, it left a lingering sense of inadequacy.

Long- and short-run objectives often conflict; sometimes the effect is merely frustrating. In this case, however, it was inspirational. The opportunity to prepare the original statement challenged me to face the subject in its broader aspects. As I went about my detailed research work, I began to conceive the idea that concerned persons in business, labor, education, the military services, and policy making persons in a larger context of public and private agencies would benefit if these ideas were worked out, however tentatively, and shared. Hence, the ideas contained in this study are, in a sense, "fall-outs" from specific project research. They are offered as a first approximation for thinking about the larger dimensions of gaps and flows of international technology.

As my ideas took shape, I became convinced that a large audience of Americans in many walks of life would also find this subject important. Other people thought so too. When I was finishing the manuscript, Servan-Schreiber's work appeared in France and I read it in the original. It was clear to me that my work was an American answer to the Frenchman's charges against the United States. My ideas of a world system of technology flowing from American initiative will be clear to readers of this book, but the contrast with the European viewpoint is worth underscoring. The Frenchman has addressed himself to the problem of what the systems analyst calls improvement of subsystem functioning; this book is global in scope. Whereas the Frenchman is concerned with a study of European problems, my concern is with the American role in the context of an international system. My book traces the alleged "Americanization of the world," at least in the technological sphere. It carries a message for people in all nations because it suggests policies for coping with international technological disruption. However, the book is addressed to Americans, and raises crucial questions of how Americans should be thinking and what they should be doing about the consequences of America's unintended spillover in the modern world.

The ideas which are sketched here of a world technology system and its components will have to be tested over the years to come, but to the extent that they provide some order in an otherwise disconnected set of elements, they are useful handles. Hopefully, such ideas will provide a scaffolding with which a more solid edifice can be built. Yet if they stimulate the public to think about these problems of international technology, to develop alternate systems of coping with these problems, and to generate informal dialogues, the mission can be said to have been accomplished.

It is impossible to acknowledge by name all the people who helped me at various stages in this study's preparation. I owe an intellectual debt to many Americans for helping me understand these problems. Europeans, British, Japanese, and people from developing countries also gave graciously of their time and thoughts. I am very grateful to Charles E. Hutchinson and William J. Price of the U.S. Air Force Office of Scientific Research for their support of the research on which much of the book was based. I wish to thank Howard S. Piquet and Professor Fritz Karl Mann for a detailed reading of the manuscript and many helpful suggestions. Acknowledgment of the valuable encouragement and assistance of Thomas F. Johnson of the American Enterprise Institution for Public Policy Research and of Dianne Littwin, Editor-in-Chief of Spartan Books is also made.

For the details of manuscript building, I wish to express my appreciation for the excellent work of my student assistants, Leon Richmond, Steve Simpson, and Bernadette Gartrell, as well as the research assistance of Annan Amegbe, instructor in the Department of Economics, Howard University, whose careful attention to bibliographical citations and other technicalities has my unstinted praise. Many thanks are also due to Mrs. Joyce Jones for final typing of the manuscript. I am heavily indebted to Miss Damaris Blosser for her painstaking editorial work, typing and correcting of numerous earlier drafts, and general managing of the manuscript. Last, but not least, I am very grateful to my wife, Flora, for her patient encouragement over a long period. However, the responsibility for any errors of commission or omission in this study is mine alone.

Washington, D.C.
January, 1970

INTRODUCTION

The subject of technology belongs to mankind, but until recently has been treated very parochially. Traditionally, knowledge of each type of technology has been the province of its specialized practitioners. With a few exceptions, notably the insightful Schumpeter and the dogmatic Marx, little attempt was made to stress the broader economic and social aspects of technology. Often technology has been confused with or overshadowed by its more distinguished sister, science. The development of modern, research-based industries has increased interest in both science and technology from the standpoint of public policy in the United States and other developed countries. However, technology often has been isolated for independent attention in order to make invidious comparisons. The term *technology gap,* which has gained wide currency, implies a narrow technological nationalism as to which nation's industry is preeminent.

Yet the new technology—its development and leads and lags from country to country—is a worldwide phenomenon which is rapidly changing the conditions of the world we inhabit, and therefore is worthy of study in a larger, more detached context. Innovations based largely on military-related research and development efforts often spill over into the civilian economy to be used for new products and practices. The absorption of these innovations often causes feelings of displacement and anxiety on the part of the individuals affected. Furthermore, the closely linked communications of the space age make such innovations known almost immediately, facilitating their transfer across international boundaries. All countries experience similar problems of adjustment to technical change. Challenge and response, rejection and absorption, lead and lag in the innovation and transfer process occur everywhere. Although the introduction of technological innovations has been going on since the dawn of history, the present systematization, acceleration, and magnification of technological change make it qualitatively different from its antecedents in today's future-oriented world. It is an epochal innovation of the modern period, as Simon Kuznets aptly terms it.[1]

Thinking persons now recognize the catalytic role of science and technology as an agent producing wealth and military power. More and more countries see the need for better educational systems as foundations for technical and scientific education, for more and better allocated research and development efforts and expenditures, for a larger supply of engineers and scientists. Their concern requires rethinking on national policies, goals, and postures—a painful process which is now going on. The subject, however, is broader than any one nation; it is international in scope.

This book is offered as a first attempt to view modern technology in an international context. It seeks to show the mechanism of world technology from its origin in the laboratory through its transfer abroad to the economic, political, and social changes it generates and the adjustments it requires. The mechanism is complex, embedded in international trade and business, in international political and social relations, and in military-related research and development efforts, but if the subject is approached with a concern for reality and a desire to get the most from the world's available technological resources, some manageable thinking emerges. While there are no simple answers, it is believed that a framework has been outlined offering the opportunity for more objective study of such popular plaints as the "technology gap," "brain drain," "technological empires," "American challenge," or "American challenge challenged." [2]

This study begins with the widely publicized technology gap, a graphic phrase reflecting this basic dissatisfaction and mankind's gnawing anxiety over the need to place in the technology race. Significantly, the developed countries first raised the cry of "technology gap"; the underdeveloped world perhaps is too far behind to perceive its need and too concerned with getting a little technology transferred to themselves to have raised the issue. But the latter must be studied also in relation to the problem. The technological world is truly one. The British scientist who is "brain-drained" to the United States is often replaced in Britain by another man from a less developed country. The web of interrelationships must be delineated and specified in order that the mechanism may be understood and utilized to mankind's advantage.

The acceleration of technological change makes it new and disturbing. Historically, technology has moved from one part of the world to another by a creeping process of diffusion. Today, technology is increasingly transferred purposively and creatively. Transfer of technology reflects the strong, rationalizing, analyzing, programming, and achieving spirit of modern men. This Faustian force of the new technology is fundamental to the future and may mean disruption of much of the system under which men have lived since the eighteenth century.

This new technology, largely born of American practice, encompasses more than new techniques, important as these are. As the introductory quotation shows, De Tocqueville, a century and a half ago, perceived that it is peculiarly American to build with the next generation of technology in mind. Likewise, the new technology is a system with application as its end product. It extends from the birth of the concept to the actualization of the final product, from its use in the weapons system to its commercial application. It is best considered as a system which includes management, finance, marketing, sales, and even postsales service follow-up.

The system originated in the United States, although nationals of other countries have contributed innovations and much of the basic science. The system evolved from America's conditions—its large-scale market, its egalitarianism, its mass education, its practical orientation of training provided by engineering and business schools, its productivity-conscious labor force supporting a flexible government infrastructure, and its consumers attuned to credit buying, quality, and efficiency. In addition, the new technology in the United States is to a considerable extent a product of heavy government funding of research and development efforts concentrated in the military-nuclear-space area.

There is no point in condemning this concentration, as some naïve persons like to do, by contrasting it with alternative dispositions of the resources. These alternatives do not exist as realistic options, nor are they likely to in a foreseeable time horizon of five or ten years. Until the likelihood of changed emphasis appears, it is helpful to recognize that the present research concentration is a fountainhead of innovations which are changing profoundly the world in which we are living. The new research complex has evolved methods to solve problems effectively and at reasonable cost, but it requires large-scale economies. Researchers think also in terms of goals, alternatives, and know how to plan and program effectively for lengthy time spans in the future. The system constantly generates unintended results in countless fields that are of great benefit to mankind. Hospital and medical technique, for example, has received an innumerable array of lifesaving and electronic aids which did not exist even a few years ago.

This study is not as concerned with describing these important spillovers as with the process of transferring them from one part of the world to another. The innovations that develop from the research and development complex move through various channels from the United States to advanced countries and underdeveloped countries. These innovations create challenges, difficulties, and tensions, but they also generate feedbacks. Others receiving the innovations begin to question policies which hitherto have worked well. Although the transfer channels themselves—government, business, and military—have historic antecedents, they are evolving in unexpected ways under the impact of fast-moving events. Recognition of the force and some of the dimensions of this new research-based technology in transfer is of major importance for all persons concerned with a better functioning world techno-economic system.

The present book is conceived in three parts. The first attempts to state the problem of modern technology—the gaps it creates and the transfers it engenders. The second part is concerned with the major participants in the world technological circuit: the United States, the developed countries, and the developing or less developed countries. The Soviet sphere is touched lightly because, involved in its own affairs, it seems to be largely outside of the present world techno-economic system. This condition, however, is changing, and perhaps in a few years the Soviets will take a more affirmative role in technological transfer.[3] But today, the giant supports of the free world's international technology transfer are grounded in the triangle of the developed economies—the United States, Europe, and Japan—and those developing countries which have been following a Western orientation. Their respective technological positions are examined in the middle chapters. The third part is concerned with channels of technological transfer, improvement, and consequences. Some key channels—government, military, and business—are analyzed from the viewpoint of improving the mechanisms for more equitable and efficient transfer of technology. The last chapters attempt to draw the strands together, and offer some alternatives for the future.

Implicit in the approach is the idea that the technological world is a universal but unitary system, and that it is to humanity's advantage to transfer the new technology as efficiently, smoothly, and expeditiously as possible. As each element in the circuit is understood more widely, dialogues are opened; the people of each country become more reasonable about their position, and their attitudes are changed by persuasion instead of forceful direction. In the last analysis, it is the trade-offs between the interfaces of any system which make it function more efficiently. Just which trade-offs may be possible depend on understanding the existing positions and preferences of the participants in the world system, and alternative channels of technological transfer and options which are open to them.

Moreover, these nice balancings of detached trade-offs are being tested with the unplanned consequences developing out of the ongoing system of transfer of technology. The decline in the U.S. trade surplus in the late 1960's exemplifies such unintended effects and will provide a test of policy. The postwar period which has seen the flowering of the American technological preeminence and thrust in the world economy is already experiencing the counterforce generated by the rising power of European and Japanese exports. This counterthrust is largely the result of the initial "American challenge," to which the response of fighting fire with fire has already begun. European and Japanese industry has already rationalized itself and borrowed or licensed American technique to the point where their capabilities in matching American technology-based exports in many areas are only too apparent. While the counterchallenge has not occurred in research-intensive fields connected with aerospace like electronics, the burgeoning imports of Volkswagen and Japanese cars are typical of many industries. It is true that the export surplus is being replaced by the income from investments abroad, but the fact remains that European and Japanese competition has indeed absorbed so much of American technology, including managerial and marketing practices, that historic American leads in many areas have turned to lags. Furthermore, political attacks on research and development under military contracts can be expected to weaken still further the U.S. position vis-à-vis world technology. As the world moves into the 1970's, more and more industries in the United States may find themselves in the position of other countries in the 1950's and 1960's. The advice then was to imitate, modernize, and meet the competition. At least some part of American industry's prospects for the 1970's would seem to depend on the traditional prescription.

NOTES TO THE INTRODUCTION

1. Simon Kuznets, *Modern Economic Growth, Rate, Structure, and Spread* (New Haven and London: Yale University Press, 1966), p. 8.

2. Jean-Jacques Servan-Schreiber, *The American Challenge* (New York: Atheneum, 1969). French edition *Le Defi Americain* (Paris: Denoel, 1967); reply: John P. Rhodes, "The American Challenge Challenged," *Harvard Business Review,* Vol. 47 (September–October, 1969), pp. 45–57.

3. In the past, the Soviets received much Western technology in transfer, see Wladimir Naleszkiewicz, "Technical Assistance of the American Enterprises to the Growth of the Soviet Union, 1929–1933," *The Russian Review,* Vol. 25, No. 1 (January, 1966), 54–76; Anthony C. Sutton, *Western Technology and Soviet Economic Development 1917 to 1930* (Palo Ato, Calif.: Stanford University Hoover Institution, 1968); also Organization for European Economic Development and Cooperation, *Science Policy in the USSR* (Paris: OECD, 1969), which predicts continuation of the still chronic "technology gap."

I

TECHNOLOGY IN THE MODERN WORLD

1
THE TECHNOLOGY GAP: MYTH AND REALITY

Nowadays, it is popular to speak of gaps. The technology gap is an outstanding member of the gap family, and like its kin it is often a thought-inhibiting rather than a thought-provoking phrase. It has impressive connotations. One visualizes the United States leading the world in the technology field with the rest of the world's countries lagging behind. It appears to be self-evident and hardly excites one to look beneath the surface for the real conditions. Yet the gap concept also carries an insidious implication that dire consequences are likely to befall those rich, heartless countries who are on the "have" side of the gap, particularly the United States, and that unless something is done, the "have-not" countries will somehow rise up and do something about the situation.

In contrast to the gleaming concept and its foreboding overtones, there is a counterpart term which is virtually unknown. This is the idea of technology transfer, or the fact that the existing technology gap is constantly matched by various mechanisms which tend to close the gap. The concept of technology transfer is as lusterless as the gap idea is sparkling, but as will be shown, the transfer idea is a much more substantive notion around which to organize one's thoughts about international problems.

DEFINITION OF THE GAP

But because the technology gap is the visible portion of the iceberg of complex international technology, the gap concept may be taken as a good handle for entering the larger subject. First, the idea must be examined with these questions: Is there in fact a technology gap? If so, between what countries or group of countries does it exist, and how can it be defined? Even more basic is the question: What is technology? Setting aside this last question for consideration in the next chapter, on the assumption that we have an intuitive understanding of the meaning of technology, the gap idea must be carefully specified. While no one disputes the fact that there is a general, across-the-board technological gap between the United States and Tanzania, for example, the gap between the United States and Great Britain or other advanced countries is not so clear. The United States itself technologically falls behind other countries in certain industries. West Germany leads in the chemical industry; Great Britain leads in VTOLS (Vertical Take-Off and Landing System airplanes); France leads in construction technology; Sweden excels in long-line electrical transmission; Japan leads the world in the manufacture of motorcycles and cameras and in shipbuilding. Moreover, technology leadership constantly fluctuates, and present leads may be quickly ceded. Great Britain's early lead in commercial nuclear power, for example, has now been whittled down. Loss of American leads has been noted.

Thus, the gap is susceptible to a multidimensional explanation with leads and lags existing in particular industries at particular times and places. However, all countries today, even the most underdeveloped, are each on a kind of technological escalator which leads them to make invidious comparisons with other countries' escalators, and the gap label is indeed convenient for urging modernization. This recognition does not obscure the fact that there are technology gaps rather than one gap existing between one preeminently technological country—the United States—and all other countries.

It must be acknowledged that for the most part the United States does have superior technology in a core of industries related to military, space, and nuclear efforts. These may not include certain aircraft and aircraft engines. The British designs of VTOLS, lightweight lift engines, and Hovercraft still rank near the top. Despite the challenge of the Soviets, the United States does create most of the leading technology in the field of computers, military electronics, space technology, and nuclear energy. These American industries are supported by government-sponsored research. According to the National Science Foundation, some 80 percent of all government research has been done in these areas.[1] The lead of these industries is clearly related in great part to the research effort expended. Other industries in which the United States leads are: petrochemicals, man-made fibers, steel (though not necessarily special steel), and pharmaceuticals. Although research is being done in these industries, the effort is not massive.

Part of the industrial lead of any country is due to the historic principle of comparative advantage. Industries forge ahead because of the factor endowments of the country. Each country has its natural endowments and acquired characteristics which tend to give it an advantage in certain lines. For the United States, the large continental market is one such advantage, plus the conscious decision to apply more resources to research as a line of action. But larger commitments of resources usually mean a larger availability of resources, or at least it is easier to commit larger absolute resources to research. Still other factors in the mix of conditions (such as American enterprise and the strongly competitive economic system) give the United States a comparative advantage.

Without exploring the implications of this principle, however, it is important to try to define the gap concept more precisely. Just as there is a plurality of gaps among industries as between nations, a number of gap concepts can be distinguished. First, there is a research gap, or, more exactly, a gap in the commitment of resources to research. The United States commits absolutely and proportionately more of its resources to research than does Western Europe. A study done for the Organization for Economic Cooperation and Development (OECD) in 1963 noted that there were more than a million persons engaged in research and development in the United States and the Union of Soviet Socialist Republics, compared to half that number in Western Europe, and that the United States spent about four times as much on research and development efforts as either Western European countries or the USSR when compared on a simple expenditure basis.[2] However, when allowance is made for the fact that the services of researchers are less costly in Western Europe than in the United States, the research gap was of the order of two or three to one. Excluding military and space research, the United States' lead was estimated to be only one and a half times greater than Western Europe, but military and space ratios ranged from 4:1 to 7:1, depending on how the calculation was made. Thus, there is a research gap, and it is heavily concentrated in the military and related fields.

However, it is important to make clear that the disparity between the United States and Western Europe in resource allocation for research and development does not imply consequent differences in economic growth.[3] In fact, there appears to be little correlation between economic growth and R & D expenditures. For example, the United States, with its huge research effort, lagged in economic growth in the 1950s. Simultaneously, Japan, with little indigenous research expenditure, experienced one of the highest growth rates in the world.

A second way of defining a technological gap was offered by the OECD research group through a technological balance of payments concept, defined as a comparison of what a country pays for technical know-how, licenses, and patents with its receipts for these items from abroad. Europe's technological balance of payments with the United States has been unfavorable, and is becoming more so. In 1965, the United States earned $614 million from other countries in royalty payments for technical know-how, patent royalties, and the like—about four and a half times the $138 million we paid others. In 1969, it was estimated at over $1 billion.[4] However, this concept of the gap does not take into account the productivity of the expenditure for foreign licensing. Even if a country spends more than it receives in terms of technological payments, the productivity of this expenditure may be great. Japan, for example, for a small outlay for transistor and other technology licenses, has realized billions of dollars in proceeds from export of these Japanese-made products. Furthermore, much technology also is transferred without cost through publications, conferences, or the movement of people. Similarly, technology also is embodied in plants, equipment, and other exports which are bought in international trade. For these and other reasons, this technological balance of payments concept is inaccurate, although it is one way to measure the technological gap. There is not much doubt that described in this somewhat narrow and unproductive way, the United States has a big lead over other countries.

A third and more fruitful way of assessing the technological gap is to define it as a management gap. This has been called an "innovational disparity." The gap does not exist in theoretical science but in the applied side of the sequence from idea to commercial practice. Europe and even the developing countries have outstanding scientists, augmented by an outpouring of creative scientific literature, shared and freely circulated in what has been termed "the international college of knowledge." [5] In contrast, while the United States has its share of able scientists, American industry excels in fast application, or the ability to move a scientific discovery from the laboratory through the development stages and into the market place. This includes the myriad details of the marketing process, as well as the after-sale care and service of the product. It embodies purposiveness combined with flexibility, a willingness to adapt to the hard problems of everyday experience and to "get the operation moving" smoothly in an amazingly short time. This ability to apply modern technology effectively has given the United States a dimensional lead over other developed countries. The American ability to manage change provides a hard-core reality to the technological gap concept.

The management gap is closely related to the research expenditure gap. With few exceptions, American management is alert to opportunities created by the research of a military or nuclear or space type, from which new products and processes are developed which result in increased sales and profits. The interaction of larger commitment of research resources and hyperactive, innovation-minded management in the United States creates a constant tension between the United States and the rest of the world. The gap is derived from this interacting condition and is, at best, clumsily defined by reference to them.

THE INTERNATIONAL CHALLENGE OF THE GAPS

However defined, the gap itself is perceived as a threat to foreign countries in several ways. First, the American export of new products challenges firms abroad. Second, there is the huge investment in new American subsidiaries abroad, sales of which are now approaching $200 billion. Even licensing and technical assistance agreements make the foreign firms feel overly dependent upon American technology. Joint ventures involving foreign investment frequently accompany technical licenses. Third, the so-called "brain drain" is alleged to draw scientists and engineers to work for American companies and research organizations for far larger salaries and fringe benefits than they can obtain at home. Actually, the brain drain may result from causes other than a technology gap. Simple difference in pay levels, for example, a purely economic consideration, may entice a man to another country. But myth as well as reality links brain drain with the technology gap. In any case, foreign people feel uneasy about what they perceive as a challenge from the United States, however it is manifested.

Exports of new products are probably not much of a challenge because the new technology is readily available through license and know-how agreements. Even for exports, however, there is the disturbing factor of ever-changing products impinging on more stable economies. While the foreign firms have opened the low-cost way of acquiring new technology through licenses for the manufacture of new products, this method carries the stigma of having to import instead of developing one's own technology. The technological balance of payments, however misleading, points up the one-sidedness of licensing and makes the deficit country feel uneasily dependent upon foreign or imported technology. This is especially the case of a proud, developed country such as Great Britain. Here, both myth and reality are involved.

More reality is found in the investment invasion under which American firms have seen profit opportunities in establishing manufacturing subsidiaries wherever growth potentials look highly favorable, as in the case of the European Common Market, or where a reliable, cheap, and productive labor force exists, as in Taiwan, Hong Kong, or South Korea, making for advantageous sources of low-cost products. While foreign countries usually welcome such American investments at first, they may later regard them as excessive Americanization, or, in the Marxist lexicon, American imperialism. This invasion problem has been solved, at least temporarily, by the United States government's imposition of capital controls on investment abroad as of January 1, 1968, an action taken for balance of payments reasons. To this point in time, however, there were many managerial advantages of this investment influx.

A vivid example of capital, research and management advantages is presented by the Texas Instruments Company. This company transferred commercial applications of military technology of transistors by setting up a plant in West Germany to manufacture plastic transistors and integrated circuits. Within a few months (November 1965 to March 1966), Texas Instruments Company set up a factory, recruited a German labor force, and started production. When the first plant reached full capacity, a second was built at Freising. West Germany is only one of a number of countries in which this company has such interest.[6]

Perhaps the least warranted evil ascribed to the gap which is felt by both developed and less developed countries is the migration of their native engineers and scientists to the United States. Although the migration has been going on for a long time, foreign observers point out that recently it has been intensified. American data indicate annual increases from 1,369 in 1949 to 12,523 in 1969, with sharp increases mainly the result of changes in the immigration laws, especially in 1965. However, no one is very certain of the numbers, as some scientists and engineers come to the United States for a year or two and then return home. But despite statistical uncertainty, foreign governments and persons believe the myth that too many of their nationals are leaving home to stay in the United States and that their rising numbers are harmful.[7]

What is popularly termed the "brain drain" is both a cause and an effect of the technological gap. Young scientists and engineers are attracted to the United States by higher salaries, better research facilities and working conditions, and advanced dynamic subject matter in their fields. This process reinforces itself, as the contributions of foreigners who have migrated to the United States perpetuate the lead the Americans have already established. All countries, developed and underdeveloped, which have invested money and resources to educate their young citizens naturally feel they are losing the fruits of their investment.

However, all of the foregoing assumes a nationalist point of view. If the problem is viewed from an international standpoint, the gaps and drains are constantly being filled by other transfers and replacements from other countries. The time lags which accompany the transfer of technology from country to country are constantly being shortened. Now, persons from the most remote, underdeveloped countries migrate to the most advanced countries to participate in technological development. When they return to their native countries and put their knowledge to use, they add great benefits. But even when they emigrate to live permanently in another country, there are benefits such as immigration remittances to the native country which must be counted.

The phenomenon of international transfers of technology must be examined with more detachment. The following chapters offer some foundations to this end. The nature of modern technology will be considered in Chapter 2 and its historic spread in Chapter 3. Although modern technology has some valid claim to being a thing *sui generis,* a contextual vision helps to order it. With the development of a larger world perspective, some of its distinctive character becomes a little more tractable to developing a framework. Myths yield to reality more quickly when the perspective is right.

NOTES TO CHAPTER 1

1. National Science Foundation, *Research and Development in Industry,* 1966. Surveys of Science Resources Series, NSF 68–20 (Washington, D.C.: 1968), p. 6, and *Reviews of Data on Science Resources,* NSF 69–12, February, 1969, p. 4. *The Economist,* March 16, 1968, p. 71, in a summary article based on OECD thinking, holds that the gap is more fundamental, particularly in industries which have developed in research done in the last fifteen years. In industries based on technological innovations from the 1920's and 1930's, Europe is holding its own. The article singles out the massive turnout of graduates in U.S. higher education as the underlying cause of the "gap." Other literature of interest as background includes: Edwin Mansfield (ed.), *Defense, Science and Public Policy* (New York: Norton, 1968); Richard R. Nelson, Merton J. Peck, and Edward D. Kalachek, *Technology, Economic Growth and Public Policy* (Washington, D.C.: Brookings Institution, 1967); Donald A. Schon, *Technology and Change* (New York: Dell Publishing, 1967); Daniel L. Spencer and Alexander Woroniak (eds.), *The Transfer of Technology to Developing Countries* (New York: Frederick A. Praeger, 1967).

2. C. Freeman and A. Young, *The Research and Development Effort in Western Europe, North America, and the Soviet Union* (Paris: Organization for Economic Cooperation and Development, 1965). This report was subtitled "An Experimental Comparison of Research Expenditures and Manpower" to stress the tentative nature of its efforts.

3. Joseph Rosa, "The Concern about the 'Technology Gap' Between the United States and Western Europe" (Washington, D.C., November, 1966), unpublished manuscript, p. 3.

4. Fees and royalties from U.S. direct investment were $1.279 billion in 1968 and payments were $0.112 billion—*Survey of Current Business,* October, 1969, p. 25.

5. William J. Price, "The R & D Organization's Fundamental Research Activity as a Window Between Science and Technology." A speech presented at the Ninth Institute on Research Administration at American University, Washington, D.C., April 20–24, 1964. AFOSR 65–0864 (Springfield, Va.: Clearing House for Federal, Scientific and Technical Information).

6. *Business Week,* October 20, 1966, p. 60; for a report of Texas Instruments doing the same in Italy, see *U.S. News and World Report,* January 19, 1970, p. 62.

7. Alessandro Silj, "Should Europe Recall Its Scientists? Some New Thoughts on the Brain Drain," *European Community* (Washington, D.C.: European Community Information Service, September, 1969) No. 127.

2
TECHNOLOGY AND ECONOMICS

Technology is as old as man—a million years or more. The chief record left by the cave man is his stone industry, reflecting the role of productive arts in the development of the hand and the brain. Stages in human history are characterized by advances in technology as in the Bronze Age and the Iron Age. The period from the fall of the Roman Empire to the eighteenth century, when earlier levels of technology were reattained, is thought of as decline or stagnation in technology. Yet despite the critical role of technology in supporting culture and civilization, it has been ignored until quite recently by most economists and many other analysts. One reason for this is that technology came down to us as customary practice which did not change much over time. Another reason is that technology (introduced earlier) has taken second place to its more glamourous relative, science. Many technologies have developed from scientific discoveries, such as the electrical industries which started with Michael Faraday's work in electromagnetism. But even here, theoretical science is accorded "higher" or more prestigious treatment. Yet although there is much blurring in the distinction between science and technology, it is the technology or the applied side which forms the gap between nations—at least it is the visible gap which is called into account.

Technology, however, is not only applied science, but it is a system which involves a bundle of other factors. For one, there are economic factors. Changes in technology involve better, more efficient ways of doing things which are usually more economical, at least over the long run, than the previous ways. Social, cultural, and psychological factors are also involved. Enterprising people and cultures oriented to change are more willing to accept innovations in product and practice than those with a "set" toward stability. In this respect it is instructive to examine the treatment accorded to technology in the discipline of economics.

ECONOMIC ANALYSIS AND TECHNOLOGY

Nowadays, technology as a concept has come to the fore in modern economic thinking on vital factors affecting the economy, its growth and development. Yet this recognition by economists came slowly. With certain exceptions, notably Marx, Veblen, and Schumpeter, economists tended to treat the technology factor as embodied in the historical troika of land, labor, and capital, which enter the black box as input and yield the economic output of society.

This threefold division (later reduced to two factors by combining land and capital) was always supplemented by the thought that there were possible factors which might be added in greater detailing of the economic model, but the original structure tended to freeze thought. Technology as a separate factor was not given independent recognition, but rather was held to be constant in the short run.

In the nineteenth century, there was good reason for this assumption because the technology stayed fairly constant or changed slowly. Even in the twentieth century, despite the ever-accelerating changes, it has been estimated that there has been, on the average, a fourteen- to twenty-year time lag between the scientific discovery and its widespread commercial application or consumer utilization.[1] In contrast, in those modern industries wedded to change—for instance, computers—it is an aphorism that whatever is being built today is already outmoded by systems on the drawing boards for manufacture four or five years from now.

The lag, long or short, is ascribable to the interweaving of technology and society. The complex of slow-moving, institutional factors, which must be adjusted to permit a new development to be created, reflects the socio-economic nature of technology. Technology is not a magical entity; it is entwined with many economic and social conditions. In the United States, it is closely related to the size of the market which dictates scale production—large firms, large plants, and large finance. The historical premium on scarce, well-trained labor made for the development of intensive capital and machine techniques in order to get the job done quickly. Other important factors that aided the development of technology in the United States included the American willingness to innovate, the absence of fixed traditions, the entrepreneurial spirit, and desire to get ahead through monetary rewards. Thus, when social scientists talk of technology, they open the door to a host of accompanying economic, social, and political factors which are complicated, slippery in concept, and hard to grasp. When dealing with current problems, the historical economic assumption that technology is constant, simply sidestepped the problems of technological change regardless of time lags between the idea and the actuality or between invention and innovation.

After World War II, the tidal wave of public interest in growth and development aroused closer examination and brought wider treatment of technology in economic analysis, since the idea of growth is by nature inherently long-run. However, even with the best of intentions, technology is hard to identify. Early attempts worked through the concept of technological change reflected in productivity gains. More recently, economists treated technology as a residual input in production function models—as that contribution to output which was unexplained by the conventional factors of labor and capital. Still more recently, there has been considerable division of opinion on how to treat this residual factor with the divergence arising between one school which believes technological progress is embodied in new physical capital, and a "disembodied" school which looks to less tangible manifestations of technology in new labor skills, education, or improvement in the general level of competence of the population of a country.

All of these considerations show that it is difficult to attribute new products to new technology alone. The historic contributing factors must be given a role in creating a product, however new it is. Since technology can only be traced in the economy by its concrete effects, such as a new machine or a labor skill, the present treatment of technology in economics is still in a kind of cul-de-sac.[2]

OPERATIONAL TECHNOLOGY. Fortunately, it is unnecessary to dwell on the subtleties of identifying technology's precise contributions to an economy's output. Technology is known to be, or at least thought to be, the dynamic force in modern economies, the catalytic agent bringing desirable change and increased wealth. Thus, millions of people feel that there is a widening gap between the developed countries and the developing nations of the world caused by the superior technology of the advanced countries. Similarly, Europeans believe that the technological gap between the United States and Europe is also widening, causing the loss of European scientists and engineers to the United States in the brain drain. In short, technology is believed to have an existence of its own as a new, vital force in this century.

Although technology may be defined as a way of producing goods and services, the term has several levels of meaning. First, it refers to the quality of combinations of men and machinery. A man with a shovel has less technical capital equipment to do a job than does a man with a steam shovel. Second, it refers to changes in the principle embodied in machines. A steam locomotive embodies different and less effective power than a diesel or electric locomotive. Transistors serve a function in electronics similar to that of vacuum tubes, but transistors are sturdier and more compact. Third, technology refers to vintages of machines, which though working on the same principle, are far more effective in later models than in the prototype. The latest generation of the computer is a faster, more productive instrument than its predecessor. Fourth but not least, technology refers to the way labor learns by doing. In the Horndal steel mill in Sweden, a two percent per year rate in the rise of labor productivity was observed, although there were no changes in capital investment. This observation, known as the "Horndal effect," is much celebrated in economics literature.[3] In spite of the last variant, technology is usually conceived as a way of doing something which we become aware of when there is a change in that way. Economists call this, "a shift in the production function," which implies that we can learn something of technology when there is a technological change.

Implicit in the foregoing rough definition is the fact that technology involves related labor and management skills as well as physical equipment. J. B. Quinn defines technology as knowledge about physical relationships systematically applied to useful purposes.[4] John J. Murphy believes it requires a systems approach.[5] Certainly, when speaking of technology, one would include an entire system of production, such as mass production, quality control, and research and development methods. Specifically, technology may mean a humble job, such as how to change a tire or to do first echelon maintenance on an automobile. Usually, however, it refers to something more complicated, such as jet engine technology. It may also mean the technique of doing research and development in some special field. Thus, Japanese researchers borrowed the techniques of aeronautical stress analysis.[6] All of these embody the idea of a common way of doing something, a system in practice and one which, to be noticed, must experience change. Hence, some people prefer the term *technical change* to technology.

Schumpeter, the dean of economists concerned with technology, evaded the whole problem of defining technology by talking about innovations. After distinguishing innovation from invention, Schumpeter pointed out that innovations could be broadly of two orders: product or process innovations, depending upon whether they were new products or new ways of producing these products. He defined the notion still more broadly by including new discoveries and new managerial techniques, but the distinction between new products and new ways of producing them has become better known. From this was derived another distinction of demand-creating innovations as opposed to cost-reducing innovations.

It is a popular idea that science provides the theoretical principles which later are embodied in practical inventions and innovations. While there is some truth in this, it is also true that many applications seem to have been developed without the support of pure science, as science and technology tend to pursue somewhat independent paths.[7] Yet there is an increasing, intentional effort to apply science to technology, a trend which has intensified since World War II.

RESEARCH AND DEVELOPMENT

Few persons agree on the exact definition of research and development (R & D). However, it is convenient to distinguish: (a) basic research, which is concerned with "pure" scientific knowledge; (b) applied research, which seeks to translate this basic finding into technology; and (c) development, which carries the technology into an innovation in the Schumpeterian sense of either a new product or a new process of manufacture. Therefore, R & D may be defined as the planned effort to relate concepts to commercial or social applications in some systematic way. The word *social* must be added because much R & D effort is devoted to some predetermined social goal, such as military defense or the abatement of water pollution.

However defined, research and development are the source of most technological change in our modern society. The lone inventor still exists—as witness the story of Xerox—and the role of the individual is still vital, as Jewkes and others point out in their book *The Sources of Invention*.[8] However, the planned search for better products and more efficient ways of producing them has become the province of the R & D process, particularly military-related research and development.

The fundamental research function is an increasingly complex, sophisticated, and delicate process. Funds are allocated for some generalized purpose or for solving some generalized problem, but it is not always certain what the outcome will be. A good example is the transistor development in Bell Telephone Laboratories, Inc. In this case, Bell Telephone managers faced the problem of greater extension of service for an expanding population. The sheer size of the equipment needed would create a housing problem. Bell Telephone Laboratories' researchers thought the solution might be in the field of solid state physics, and eventually William Shockley and his colleagues found the solution based on prior scientific work done at Bell and at Purdue University.[9]

The roles of industry manager and scientist in the transistor story date it. These roles used to be separate, but nowadays the scientists themselves are often found in managerial posts. Many large companies which are research-oriented are choosing scientists for officers, managers, and directors. This is true also in government; for one, Harold Brown, a recent Secretary of the United States Air Force, formerly was its chief scientist.

Scientists and researchers nowadays are likely to be practical men. University professors now serve as consultants to research projects. There is a continuous exchange from universities to industries of consultants in various specialties, and the exchange benefits both the academic and business worlds. Scientists taking part in the direction of corporations and government agencies often decide what research shall be done, or are important contributors to the decision. Scientists also hire other scientists to work with them, causing increasing demand for scientific men. The American Institute of Physics has estimated that 20,000 physicists will be required in the United States during the next ten years.[10]

As the scientist works more closely with industry, he is increasingly pressured to produce a steady flow of innovations and applications. Much of the research in modern industry in the United States is designed to improve products—to eliminate the flaws found in existing products and practices. Americans seem to feel that a higher proportion of resources spent on development rather than basic research is the correct balance.

Many basic inventions and scientific discoveries are imported from abroad. Sir Alexander Fleming, a British scientist, discovered penicillin in 1928. The Britisher B. N. Wallis invented the variable wing, although it is being developed by Boeing Aircraft Corporation to fulfill the Supersonic Transport contract. On the other hand, Dr. R. H. Goddard, an American professor of physics, was the first to use liquid propellents in rocketry, but the development of rocketry occurred in Germany. In general, Americans excel in developing the applied uses of scientific research.

NEW TECHNOLOGY EMANATING FROM MILITARY AND RELATED RESEARCH

In current usage, new, science-based technology usually refers to that which has come into being after World War II. Although it is not necessary to agree with Marshall McLuhan that the impact of the new technology has created a new communications environment for the return to tribal man and his feelings,[11] it is clear that the new technology, based largely on the research and development process, is a tremendous force which is reshaping the lives of modern men. Some dimensions of the phenomenon may be examined to set the stage for discussing the problems of the transfer of technology to all parts of the world.

The new technology also means technology which has come into being as a result of planned efforts in research and development conducted by industry under contract with the United States or other governments. Research funding by other governments is relatively small. For most purposes, United States government contracts for research, development, and production are at the center of the dynamo of technological change. Although lone inventors still originate excellent inventions which enter the innovational process, the systematized research and development process is the source of many of the startling technical changes which are transforming the twentieth century.

This research is largely military-based or military-connected. Although nuclear and space research activities may be considered separately from the military, the distinction is tenuous. The term *military research and development* can be used to subsume military, space, and nuclear activities, and as such it represents the lion's share of the research funds expended.

For objective analysis, there is nothing condemnatory in this statement; it is a fact of modern existence. Moreover, it may be observed that those who would divert these expenditures to more peaceful uses are engaging in wishful thinking, because the United States Congress as presently constituted would not appropriate money for alternative research activities on a comparable scale in the foreseeable future.

In the meantime, it is realistic to recognize the many benefits that accrue to the civilian economy through the spillover or fall-out effects from military research, and to accept the benefits and liabilities as data to study, analyze, and improve.[12] The problem of the international transfer of new technology is one of the most important of these facets which must be studied, as it affects the whole world with startling speed and in ways which were undreamed of just a few years ago.

With particular reference to the theme of this book, the gap between the United States and other advanced countries is most apparent in these newer military-research-based developments. Europe lags behind the U.S. in such growth industries as computers, electronics, and space industries, though the industries of an older vintage are still generally competitive or surpass their American counterparts. These gaps are being filled continually by international transfers of technology. The increasing ability of foreign industry to master transferred technology and to produce competitively is attested by the declining United States trade surplus in recent years. Overall, American technological leadership will continue for the foreseeable future, but many American industries are experiencing growing pressure from European and Japanese products both in the United States itself and in world export markets.[13] Such pressures stem largely from American policies of liberal *transfer of technology*. This comparatively unknown phrase, however, requires explanation as a key concept for analysis of the modern world's technological system.

Before proceeding, however, mention of the notorious unfavorable effects of technology may be made. The technology peril has been widely proclaimed; for example, this recent statement from the House Committee on Science and Astronautics:[14]

> . . . the specter of thermonuclear destruction, the tensions of congested cities, the hazards of a polluted and despoiled biosphere, the expanding arsenal of techniques for the surveillance and manipulation of private thought and behavior, the alienation of those who feel excluded from power in an increasingly technical civilization.

Such negative strictures on the effects of technological change have been repeated ad nauseum. To the end that we may count our blessings a little more carefully, a break with such carping criticism is offered in this book. It is concerned with the positive and beneficial effects of modern technology and its transfer from country to country to fill technology gaps. If neglect of the critical tradition seems important to the reader, he is referred to the copious literature of gloom and doom. The positive problem of how to transfer technology effectively and efficiently is the concern of the succeeding chapters.

NOTES TO CHAPTER 2

1. U.S. National Commission on Technology, Automation, and Economic Progress, *Technology and the American Economy* (Washington, D.C.: U.S. Government Printing Office, 1966), Vol. 1, p. 4.

2. For a more detailed discussion of the present treatment of technology in modern economic technical literature, see Daniel L. Spencer and Alexander Woroniak, "The Feasibility of Developing Transfer of Technology Functions," *Kyklos,* XX, Fasc. 2 (Basel, 1967), 431–59.

3. Ingvar Svennilson, "Economic Growth and Technical Progress: An Essay on Sequence Analysis," *The Residual Factor and Economic Growth* (Paris: Organization for Economic Cooperation and Development, 1964), pp. 103–31.

4. U.S. Department of Commerce, *Technology and World Trade,* (Washington, D.C.: 1967), p. 98.

5. John J. Murphy, "The Transfer of Technology: Retrospect and Prospect," *Transfer of Technology to Developing Nations,* eds. Daniel L. Spencer and Alexander Woroniak (New York: Frederick A. Praeger, 1967), pp. 6–9.

6. Daniel L. Spencer, *Military Transfer of Technology,* (Washington, D.C.: U.S. Air Force Office of Scientific Research, 1967), Technical Report No. 3, AD–660–537, pp. 101–03.

7. Melvin Kranzberg and Carroll W. Pursell, Jr., *Technology in Western Civilization* (New York: Oxford University Press, 1967), Vol. 1, p. 6.

8. John Jewkes, David Sawers, and Richard Stillerman, *The Sources of Invention* (London: Macmillan and Co., Ltd., 1958).

9. Richard R. Nelson, "The Link Between Science and Invention: The Case of the Transistor," *The Rate and Direction of Inventive Activity: Economic and Social Factors,* A Report of the National Bureau of Economic Research, New York (Princeton: Princeton University Press, 1962), pp. 558–83.

10. As reported in *The Times* (London), February 1, 1967, pp. 12–13. The National Science Foundation projected a 30,000 increase between 1960 and 1970, National Science Foundation, *Scientists, Engineers, and Technicians in the 1960's, NSF* 63–34 (Washington, D.C.: 1964), p. 8.

11. Marshall McLuhan, *The Gutenberg Galaxy* (Toronto, Canada: University of Toronto Press, 1962); and Marshall McLuhan, *Understanding Media* (New York: McGraw-Hill, 1964).

12. For a cautious statement calling for further research on the transferability of military-related R & D, see R. R. Rosenbloom, *Technology Transfer—Process and Policy* (Washington, D.C.: National Planning Association, 1965).

13. See above, Introduction; also, *Business Week,* December 2, 1969, pp. 202–06.

14. U.S. Congress, House of Representatives, Committee on Science and Astronautics, *Technology: Processes of Assessment and Choice* (Washington, D.C.: U.S. Government Printing Office, 1969). This report urges the establishment of a federal agency to alert the nation to perils of uncontrolled technology. Summary, *New York Times,* August 31, 1969, p. 28.

3
THE INTERNATIONAL TRANSFER OF TECHNOLOGY

The transfer of technology around the world is as important as its development. Historically, technology and its new products were diffused in what may be called an unplanned, natural manner. Like every other country, the United States also is deeply indebted to this historic diffusion. The following quotation from Ralph Linton serves to remind Americans of this debt:[1]

> Our solid American citizen awakens in a bed built on a pattern which originated in the Near East but which was modified in northern Europe before it was transmitted to America. He throws back covers made from cotton, domesticated in India, or linen, domesticated in the Near East, or wool from a sheep, also domesticated in the Near East, or silk, the use of which was discovered in China. All of these materials have been spun and woven by processes invented in the Near East. He slips into his moccasins, invented by the Indians of the Eastern woodlands, and goes to the bathroom, whose fixtures are a mixture of European and American inventions, both of recent date. He takes off his pajamas, a garment invented in India, and washes with soap invented by the ancient Gauls. He then shaves, a masochistic rite which seems to have been derived from either Sumer or ancient Egypt. . . .
>
> On his way to breakfast, he stops to buy a paper, paying for it with coins, an ancient Lydian invention. At the restaurant a whole new series of borrowed elements confronts him. His plate is made of a form of pottery invented in China. His knife is of steel, an alloy first made in southern India, his fork, a medieval Italian invention, and his spoon, a derivative of a Roman original. He begins breakfast with an orange, from the eastern Mediterranean, a cantaloupe from Persia, or perhaps a piece of

> African watermelon. With this he has coffee, an Abyssinian plant, with cream and sugar. Both the domestication of cows and idea of milking them originated in the Near East, while sugar was first made in India. After his fruit and first coffee he goes on to waffles, cakes made by a Scandinavian technique from wheat domesticated in Asia Minor. Over these he pours maple syrup, invented by the Indians in the Eastern woodlands. As a side dish he may have the egg of a species of bird domesticated in Indo-China, or thin strips of the flesh of an animal domesticated in Eastern Asia which have been salted and smoked by a process developed in northern Europe.
>
> When our friend has finished eating, he settles back to smoke, an American Indian habit, consuming a plant domesticated in Brazil in either a pipe, derived from the Indians of Virginia, or a cigarette, derived from Mexico. If he is hardy enough, he may even attempt a cigar, transmitted to us from the Antilles by way of Spain. While smoking he reads the news of the day, imprinted in characters invented by the ancient Semites upon a material invented in China by a process invented in Germany. As he absorbs the accounts of foreign troubles, he will, if he is a good conservative citizen, thank a Hebrew deity in an Indo-European language that he is 100 percent American.

This historic diffusion process still continues. Sooner or later, ideas, products, and technologies spread, are absorbed into the recipient system, and are quickly taken for granted as indigenous.

However, the modern age has been characterized by a somewhat different phenomenon, namely, the transfer of technology which is a planned, predetermined effort to make technology available to those who lack it. It is the counterpart concept of the technological gap—it is the effort to close that gap. This process goes on constantly, through different channels, but all directed at reducing the gap. The gap is plural and fluid; as fast as some part of it is closed, the gap widens elsewhere, but this does not change the transfer process. It suggests instead that technology transfer is likely to be a continuing study for the foreseeable future.

TECHNOLOGY TRANSFER

What are some of the characteristics of this key concept, transfer of technology? How does the mechanism operate? Who undertakes it and under what circumstances and conditions? What makes for successful transfer? These are some of the typical questions which are asked about this promising new tool of thought which is becoming increasingly popular in solving modern problems of change.

Literature is still not extensive in the field, but a recent bibliography lists close to 500 references on this subject.[2]

Basically, the term means a planned and rational movement of information and technique on how to perform some task, simple or complex. It should be distinguished from diffusion, the historic, unplanned movements of technical or social items from one country to another or from one user to another in the same system. Historic diffusion just happened; movement of such items as in the quotation from Linton was not planned and was the product of infinitely simpler technologies. It was generally a slow process—like the ooze of liquid seepage. Imperceptibly, usually over long periods of time, innovations oozed from place to place. This historic ooze was largely unplanned, unpremeditated, and barely noticed. In contrast, the present-day technology transfer results from some carefully considered planning and programming for controlled action, with feedback and monitoring of the success or failure of the results. Fortuitious circumstances, serendipity, of course, result in unexpected discoveries and spillovers, but these play a comparatively small role in innovation and technology transfer. Typically, there is a demand to fill some gap, and an orderly search is started to transfer technology into the breech. Transfer of technology is essentially a rational concept, resulting in rational procedures and actions, although it is fashionable to emphasize irrational elements in some literature of innovation and research.[3]

Transfer of technology in its simplest formulation refers to the purposive movement of established technology or technical innovation (incremental technology or new products) from place to place, company to company, or use to use. Thus, technology is transferred by one company to another through a license, exchange of patents, or a know-how agreement. It may be a transfer from Illinois to California, or it may be from an American company to a foreign subsidiary or foreign corporation. It usually involves economic application, but it also may not, as in some knowledge-oriented transactions; for example, when a foreign student learns space technology which may not be commercially exploitable—at least in the short run. It may be a transfer of technology from one use to another, for example, in the case of the bearings on the fast train between Tokyo and Osaka, which were transferred from technology borrowed under license by Japanese aircraft companies and transferred again to another use.

The National Aeronautics and Space Agency (NASA) and the Atomic Energy Commission (AEC) have developed elaborate programs to organize, retrieve, and disseminate technical information. The NASA program, to which AEC was similar, consists of six regional dissemination centers established to make scientific and technical information available to business and other organizations. The information consists of references to documents in the NASA data base stored on tapes and microfiche copies of reports on technology developed in the space program. Technical briefs and back-up publications are made available as well as personalized communications. According to the agency, 11 million copies of 3,000 technical briefs have been distributed. Over the past eight years, results have been slow, but some success is reported, particularly in education. For example, technical utilization of NASA data is being developed to reduce the lag in textbook development. In industry, a follow-up study of 11,000 cases of technical support packages revealed that while only 1 percent resulted directly in new products, 44 percent of the cases studied reported some value to the firm.[4] Other areas for transfer are now being explored, for example, developing countries, environment, and urban problems.[5] The agency looks forward to an accretion of effort resulting in a critical mass effect as the transfer cumulates over the years and a threshold of large scale penetration is reached.

All of the above presumes a gap and some premeditated effort to transfer technology to fill the gap. A more complex vision of the transfer process comes about when technology in one field becomes available, and observant entrepreneurs or alert individuals in other fields realize that it can be used for some hitherto unknown product or previously unfelt need. "TV dinners" were transferred from commercial airlines (previously direct transfers from military aircraft research and development) to a "created need" for quick consumer dinners. Other sophisticated models of transfer of technology have been developed, including vertical transfer in the science vis-a-vis technology interface.[6] These elaborations are interesting but marginal to the main line usage, of which international transfer of technology is a part. Normally, a need or gap stimulates transfer of technology, and the more relevant problem is how to effect it expeditiously and efficiently.

In industry the methods and priorities are described by one research administrator as follows:[7] First, industry tries to hire the man who developed or knows the technology. Failing this, the next best thing to do is to try to exchange the know-how for proprietary information on another technology which the company has developed. Third, the company will try to buy or license the technology; and only as a last resort will it try to develop the technology independently. Principles of transfer of technology in the public sector in such fields as education, medicine, urban environment, and other economic areas may be somewhat different from industry, but the order of priorities in seeking to effect the transfer is suggestive.

Rogers, in a classical study of the movement of innovations drawn from many fields, has summarized the process as having five stages: awareness, interest, evaluation, trial, and adoption.[8] His illustration of hybrid seed describes the typical sequence. A farmer first learned of the seed in a farm magazine in 1933, but it was not until 1936 that he bought and planted a trial bushel. Trial and adoption take place after this and later tests of the innovation. On an average, farmers take nine years to go through the process. Mansfield sets forth some propositions which govern the decision to accept the transfer: the profitability of the innovation, the obstacles to transfer and adoption, and the risks involved in changing from the old method.[9] The prestige and influence of the auspices under which the transfer of technology takes place is important in acceptance and rate of adoption.

This consideration brings up another dimension of the subject: the change agent, or who it is that undertakes technology transfer. It was formerly thought that dissemination of technical literature was an efficient mechanism for the task of technology transfer. While the mass accretion of technical information dissemination is still a promising method for specific transfers to take place, product champions are often necessary; that is, in many cases, there must be committed, dedicated individuals willing to commit themselves to sacrifices of position and prestige to transfer the innovation into an ongoing system. Rogers noted that any innovator must be a deviant personality, but that technology transfer is best accomplished by leaders or prestigious individuals or firms or agencies adopting the innovation initially. These leaders are usually cosmopolitan, adventuresome persons who seek out and try new things. The Japanese, famed for their speed of acquisition of new technologies, have cultivated an entire system of seeking out and transferring foreign innovations. In Japan, the man who goes abroad to watch foreign technological developments is known as an "antenna"; the system may be called the antenna system. These antennas are business representatives in their specialties who constantly seek innovations and refer them back to research and development men in their companies at home.[10]

Entrepreneurs, especially in small companies, are also key persons in the technology transfer process. Route 128 outside of Boston has become symbolic of the highly competent small R & D firms, often developed as spin-offs of some commercial aspects of research projects at university laboratories under government contracts. At present, they typically provide interdisciplinary teams who bid for and carry out contract projects in such areas as nuclear energy, weapons systems, space, and more recently in such areas as oceanography, environmental pollution, and problems of the inner city. These small talented companies are likely to become increasingly important in the transfer of technology to urban and social problems.[11] They are in a sense go-betweens or marriage makers between the needs and the existent technology. The proliferation of these small "broker" firms at the interface between government and private activity testifies to the importance of the entrepreneur and the vitality of competitive market activity in technology transfer. Within organizations Allen has noted the importance of "technological gatekeepers," or key individuals whose command of several "informational coding schemes" may break down barriers of communication and make possible effective technology transfer across previous barriers.[12]

The importance of the human agent in the transfer of technology has even been carried to the point that one observer calls for a new model in which transfer of technology is defined simply as the transfer of a person. The traditional idea of transfer of technology as the movement of knowledge in "an assemblage of pieces of information which can be extracted or expelled from one sector of organized creativity and transposed to another to produce different outputs" is wrong because it does not fit modern conditions, according to Burns.[13] However, most people, while recognizing the importance of the human agent, would define transfer of technology as the transfer of ideas and practices couched in license, drawings, manuals, know-how contracts, training, etc. But Burns' extreme view underscores the catalytic role of the individual in transfer of technology.

International transfer of technology, the concern of this book, is a branch of the general subject; and as such, the principles do not differ markedly from the domestic side. However, some differences are apparent, perhaps more in degree than in kind. The human factor may be even more important and the barriers perhaps a little higher because of nationalism, lack of homogeneity, and traditional, nonmodern values. A country whose leaders feel in some way demeaned by borrowing technology developed in another country suffers an additional impedance in plugging into the circuits of modern technology. A country is burdened when its elites lack a sense of emergency or when its people will not follow instructions for lack of training, morality, or lack of identification with the leadership. A country with massive divisive forces of caste, class, tribe, tradition, and political problems is diverted from concentrating on the specialized problems of international transfer of technology. A great handicap for a people is the lack of achievement orientation, or its loss by a people grown too comfortable for acceptance of innovation and change. Conversely, there are advantages in the international arena. A people like the Japanese, with a strong sense of national identity, a need to promote the common good, and a vital tradition of enterprise and achievement, have advantages. Japan's propensity to borrow and integrate foreign innovations to the system quickly and efficiently is a remarkable model, which is popularly derided as copying, but might better be emulated as a model of catching up.

CHANNELS OF INTERNATIONAL TRANSFER

The channels of technology are hard to classify, and at best they overlap. The first and most obvious channel is the government or, more exactly, government to government. There are also the efforts of the United Nations and related agencies like the World Bank which can be termed international government bodies. This government activity is often thought of as making finances available for projects under economic aid programs. However, these fundings conceal a heavy technical component which is transferred. Economic aid is often distinguished from technical assistance programs which are part of planned technology transfer. Either technicians of the developed country go to the underdeveloped country to teach, or nationals of the recipient country come to the donor nation for training. This may either be the long-term training of engineers or scientists in the universities, or short-run, follow-up training for men who are already graduates. The productivity team program dating from the days of the Marshall Plan illustrates the latter type. Military assistance is another of the government channels, but it is so important, far-reaching, and yet little known that it requires a separate discussion, which is provided in Chapter 9.

Next comes the business channel through which the most complicated technologies are transferred. In considering the business channels, it is convenient to use Professor Ingvar Svennilson's distinction between "technical knowledge" and "know-how."[14] Technical knowledge is the wider, more precise concept embracing engineering and managerial marketing or distributional activities necessary to obtain a specific output. "Know-how," in contrast, adds to textbook knowledge the skill of the application in actual practice. A young man from an underdeveloped country may study engineering in the advanced country, obtaining with relative ease a textbook understanding of his field, yet he may find it difficult to convert this knowledge into the know-how of the industrial plant in his own country. "Know-how" often involves innovations which are patented business secrets of a firm or system secrets of the industry.

Much of thte technology of the know-how type is transferred through business channels by licensing and know-how agreements. These transfers are acquired by the payment of royalties and other fees and are relatively cheap, if one considers the alternative cost of developing the same technology in the recipient country. It is also an effective way of keeping up to date in advanced technology and is a specific antidote to the technology gap. The Japanese, from whom few cries of "technological gap" are heard, have a saying that anything developed in the United States economy will be in Japan six months later.

Direct investment in a country by a foreign company is another direct channel of transfer of technology. As will be shown later, this channel suffers from conflict of interest, supposed or real, between the country and the investing foreign firm. In spite of this conflict, free and easy communication between a parent firm and subsidiary is one of the most effective methods of transfer of technology. It provides a package of capital, skills, and managerial talent which can inject an effective transfer into even a backward and hostile environment.

Lastly, there are person-to-person transfers of technology; one of the simplest and most readily available is the reading of technical and scientific journals and reports. These are available for a small subscription cost and sometimes for nothing. Companies often make available to their licensees reprints of reports by their research departments. These may or may not be utilized, depending upon the ability of the licensee to absorb such findings.

Another transfer mechanism is that of the university. Students from less developed countries come to study in the universities of developed countries, which in turn also send teachers to universities and schools of the less developed nations. A whole new dimension has been added in relationships between institutions of higher learning in developed and developing countries. *Institution-building* is the term sometimes applied to this type of transfer. Often these students and learning centers are government-financed, and thus overlap with government channels of transfer. There are also many privately financed transfers through the medium of education.

CONDITIONS FOR TRANSFER

To be effective, these channels of transfer must be subject to certain conditions. A less developed nation which wishes to benefit from the transfer of technology must have sound government policies. A government cannot use scarce resources to build palaces and other showy projects when it fails to finance its ablest youth to study abroad. As to the resources themselves, the nation must have an economy effective enough to generate the income, and the foreign exchange to finance students' education, to finance technology licensing, and to obtain loans and aid from foreign bankers and governments. In addition, the people of the less developed country must have the flexibility to accept foreign ideas into their own economic system. The propensity to borrow foreign technology is one of the most important factors pertaining to a less developed country's attempts to advance technologically. Japan has shown remarkable willingness to accept foreign technology, and the success of her great modern industries is based generally on simple copying of foreign innovations.[15]

In contrast, other nations have been remarkably hesitant to imitate even a manifestly better practice. England, due to her lack of response to scientific and technical change, in many areas early lost her lead to less inhibited competition, notably the United States and Germany.[16]

Another principal condition for successful transfer is the willingness of the advanced country to make its technology available. Industries in advanced countries do not welcome new competition, even though they pay lip service to the principle. Nevertheless, some nations have been relatively generous in making their technology available. The furor in Great Britain over the proposal to increase the university fees charged foreign students points up the importance of the supplier side. In the past, the United States has exhibited a relative openhandedness in supplying its technology through the channels of student migration, postgraduate work through on-the-job training in American industry and research centers, productivity teams, or openhanded licensing arrangements which can be had almost for the asking. While there are some signs that American business is more cautious as gaps have closed, technology will continue to be licensed, though perhaps on stricter terms.

What are the benefits to the donor in return for extending assistance? On a political level, the donor country has the assurance that it is not losing contact with poorer countries; on the socio-economic level, its system is being approached to some degree, and it avoids isolation while building up other modern peoples with like-minded interests. Basically, the *quid pro quo* is an exchange of technology in return for military alliances which counter the spread of communism. However, altruism may also motivate many exchanges.

On the economic level, the donor country's commercial/industrial firms obtain profits and other business advantages, such as forestalling competition. Markets overseas often offer opportunities for more rapid growth potentials than the home market. For the United States firm, a product developed as a spin-off from military research and development has, for a time, monopolistic advantages. In an overseas market, the new product may have greater sales potential than at home. Tariff barriers and other restrictions may force a company of one country to gain entry into another by setting up a new plant or licensing an affiliate in the restricted country. Cheap labor or other local supply advantages in one country may offer special attractions to a foreign company to make its technology available through investment or licensing. While these are just a few of the factors,[17] they serve to show the complex motivations for the transfer of technology from one country to another.

There are similar complex motivations on the recipient's side. The government, business interests, elite, and the people as a whole of the donee nation want to be modern. It is an overused yet basic thought that the countries of the world have been brought closer together by television, radio, and jet airplane communications. New developments in technology are quickly reported to the world, and the demonstration effect occurs on the motion picture or television screen.[18] The peoples of less developed countries soon become aware of the lags and gaps in their own technologies and economies. The peoples who have less are inspired to improve their living conditions, but the efforts, costs, and sacrifices of making necessary changes are temporarily obscured.

Each country holds some position on the technological world scale of today, and each country's industries can be compared to the same type in other countries. Thus, each country is constantly making invidious comparisons, although much of this technological nationalism is really unnecessary. The material in this book has been assembled to show that there is a world technological system which, like the world trading system to which it is closely related, has benefits for all participating countries. The second part of this book attempts to describe the polarities of the system, and suggests that intelligent trade-offs between the interests, aspirations, and availabilities of these components may be the best prospect for adjustment to technological change in the remaining years of the twentieth century.

NOTES TO CHAPTER 3

1. Ralph Linton, *The Study of Man* (New York and London: D. Appleton-Century Co., 1936), pp. 326–27.

2. M. Terry Sovel, *Technology Transfer—A Selected Bibliography* (Denver, Colo.: University of Denver, 1969), NASA CR–1355.

3. The irrational theme appears, inter alia, in Donald Schon, *Technology and Change* (New York: Dell Publishing, 1967), Ch. 1, passim; Tom Burns, "Models, Images, and Myths," *Factors in the Transfer of Technology*, William H. Gruber and Donald G. Marquis (eds.) (Cambridge, Mass.: The Massachusetts Institute of Technology Press, 1969). In passing, it may be noted that the operation of modern techno-industrial systems, of which innovation and transfer of technology are important elements, was hardly built on Hamlet-style doubts by snide intellectuals.

4. Denver Research Institute, *Project for the Analysis of Technology Transfer—The Initial Year*, prepared for NASA (Denver, Colo.: Industrial Economics Division, Denver Research Institute, November, 1968), p. 39.

5. Peter E. Glazer, *et al., Space Technology Transfer and Developing Nations* (Cambridge, Mass.: Arthur D. Littel, 1968), NASA CR–1222; also footnote 11 below.

6. Gruber and Marquis, *op. cit.*, pp. 4–8.

7. Egon Loebner, Hewlett-Packard Corporation, at Denver Research Institute, *Conference on the Environment and the Action in the Transfer of Technology*, Snowmass-at-Aspen, Colorado, September 26–28, 1969.

8. Everett M. Rogers, *Diffusion of Innovations* (New York: Macmillan, 1962).

9. Edwin Mansfield, *Economics of Technological Change* (New York: Norton, 1967), pp. 119–20.

10. See Daniel L. Spencer's research, discussed and cited in Chapters 7 and 9.

11. For an excellent analysis of the use of systems technique in the urban problems, see Robert Chartrand, *et al., Systems Technology Applied to Social and Community Problems,* U.S. Senate, Committee on Labor and Public Welfare, Subcommittee on Employment, Manpower, and Poverty (Washington, D.C.: U.S. Government Printing Office, 1969; New York: Spartan Books, 1970).

12. Thomas J. Allen, "The Differential Performance of Information Channels in the Transfer of Technology," *Factors in the Transfer of Technology, op. cit.,* p. 153.

13. Burns, *op. cit.*

14. Ingvar Svennilson, "Transfer of Industrial Know-How to Nonindustrial Countries," *Economic Development: with Special Reference to East Asia,* Kenneth Berrill, (ed.) (New York: St. Martin's Press, 1964), pp. 405–28.

15. Daniel L. Spencer, *Military Transfer of Technology,* (Washington, D.C.: U.S. Air Force Office of Scientific Research), Technical Report No 3, AD–660–537, p. 108. Also Japanese industrialists are quoted as taking the motto: "Find out the best practice in the world and improve on that," in "Two Island Nations: A Study in Contrasts," *U.S. News and World Report,* January 16, 1967, p. 59; Daniel L. Spencer, "New Technology in Japan," *World Affairs,* Vol. 132 (June, 1969), pp. 13–28.

16. D. H. Aldcroft, "The Entrepreneur and the British Economy, 1870–1914," *Economic History Review,* Second Series, XVII (August, 1964), 113–34.

17. For a further discussion of the advantages to United States businesses, see National Industrial Conference Board, *Foreign Licensing Agreements* (New York: 1959).

18. "Demonstration Effect" has become famous in literature in the development field, e.g., Charles P. Kindleberger, *Economic Development* (2nd ed. New York: McGraw-Hill, 1965), pp. 141–43.

II

POLARITIES OF THE INTERNATIONAL TECHNOLOGICAL SYSTEM

4

THE UNITED STATES: POSITION AND POLICY

After World War II, the United States emerged both as the central magnet and as the generator for international transfer of technology. However, American technological leadership may well be dated by Henry Ford's innovation of assembly lines in automobile production before World War I. During the 1920's, the character of American technical civilization was identifiable with mass production, mass consumption, and a steady flow of innovations. Technological gains were consolidated during the business depression of the 1930's, but fresh successes were achieved during World War II, when the American technological system scored with its vast outpouring of war materiel production, with its in-depth logistical system, and with the swift, efficient management of a supply line that extended thousands of miles to serve war needs on many fronts. However, despite this solid basis, the dimensional difference between the technology of the United States and that of the other developed countries has become most apparent since 1945.

Following World War II, the United States government put into operation the famed Marshall Plan to help rebuild European countries suffering from the war's destruction, and U.S. aid and procurements also helped to rebuild Japan's economy. This period also revealed that many British, European, and Japanese technologies were outmoded. American technicians and managers went overseas to assist in reconstruction and development of the economies of other countries. American technology also was made available to the visiting productivity teams from other nations who poured into the United States. Massive infusions of American grant aid also accompanied the transfer of American technology to other countries. As a result, the world's developed countries made spectacular recoveries, and during the 1950's they returned to a prewar equilibrium with the United States—even surpassing it in economic growth rates. The United States and the USSR were recognized as superstates with vast military and political power, but Western Europe brought forth the Common Market to offset some of the differences in market size. Technology was not visualized as a gap or a threat until the 1960's.

THE INTERNATIONAL TECHNOLOGY POLICY OF THE UNITED STATES

If it could be said to have a policy on technology, the United States government viewed technology as part of its trade policy. Technology was embodied in the factor endowment of a country which trades its goods in the international economy. Alternatively, technology was a product or process or method embodied in a patent or a set of drawings which were commercial objects to be traded in the international market. As in the domestic market, the best policy for nations was free or freer trade, and insofar as technology was a tangible item, it was to be bought and sold in free markets with as little regulation as possible. National security considerations might provide an exception to such rules, and therefore restrictions on trade with the Soviet bloc were established. Technology often was made available to allies on the express condition that they did not send the technology behind the Iron Curtain.[1]

The free trade principle for trade in goods also was extended to the factors of production. The desirable condition was as little restraint as possible on the movement of capital. Free movement of capital was seen as moving primarily in response to differentials in return, either in the long or short run. A desirable equilibrium would come about with the capital-rich nations making available their capital to those areas where scarcity dictated a higher return on investment. All men would tend to benefit as poorer nations obtained more income and became better trading partners for the capital-providing nations, which in turn would become still more prosperous. Technology was also included, embodied in successive waves of capital installations.

As with capital, so with labor, skilled or unskilled. The international migration of people had the same equalizing effects which would make for a more efficient, prosperous world over the long run. Unskilled labor migrations flowed into unpopulated areas and made possible the development of their resources and their addition to the international trading network. Skilled labor spread the benefits of technology and science on a person-to-person basis. During the eighteenth and nineteenth centuries, beneficial migrations of both types of labor took place. The United States has always been conscious that its own heritage was derived from these fundamental migrations of unskilled and skilled labor.

Underlying these ideas was the fundamental tenet that there was freedom of the individual to make his own choices as to the disposition of his life and property. In theory at least, the freedom of the individual transcended even the needs of the national state. Anything which interfered with the freedom of choice of individuals to seek new opportunities, by extension, interfered with the free choice of individuals to share experiences in the international community of science and technology.

These individualistic ideas have been tempered in the United States, as in all countries, by the needs of the national entity, and modified by the demands of other important goals. Protectionist theories, supported by pressure groups inherent in the representative government structure of a modern democracy, have countered the individualistic theory in many respects. The free movement of goods in international trade conflicts to some degree with the products of domestic industry, with domestic employment, and, above all, with national economic development. During the first years of the United States as a republic, Alexander Hamilton argued for protection of new industries, and in modern times the protectionist position has been tied to the economic development argument which holds that industries, now inefficient, will become more efficient with time. Free trade, the protectionists argue, is desirable among states with a similar structural development, but one-sided trade between industrialized nations and weak, unintegrated, monocrop economies disrupts and exploits the weaker countries.

Similarly, the flow of the productive ingredients, capital and labor, can have demoralizing and destructive effects. As long as Marxian dogmas of exploitation and imperialism are endlessly proclaimed, with little reality in a world of socially conscious, international business, the introduction of foreign capital into a less developed country will be resented. The local population feels strongly about the sharp dualism of their static, agriculturally based economy on the one side and the vigorous, modern, foreign capital enclave on the other. Beneficial economic effects for the poor country derived from the investment of foreign capital are offset in the minds of the people by the feeling that rich outsiders are taking over their economy, influencing their government, and somehow reducing them to an inferior status. To counteract this hostility, joint ventures and other forms of mutual participation have been introduced in recent years, but such countermeasures have not completely dispelled the suspicions of peoples in the poor nations. American policy is tempered accordingly.

Regarding the international migration of peoples, the United States like other nations until recently followed a selective immigration policy. Yet the United States has been more willing than most nations to accept the foreign immigrant into its system, to study in its schools, to stay for postgraduate, on-the-job training, and finally to remain as a resident or citizen if the individual's skills or personal circumstances accord with defined national policies.[2]

The United States and other countries accepting immigrants assumed that the act of emigrating indicated that these individuals were surplus factors in the society from which they came and that the immigrants' talents and labor enriched the countries receiving them. There was a certain validity in this position when only a few educated, well-trained individuals emigrated to other countries. The recent worldwide concern with the brain drain indicates that, at least in popular belief, both developed and developing countries are losing more of their scientists and technical elite to the United States at an accelerated rate.

INTERNATIONAL BUSINESS AND THE R & D EXPLOSION

The new technology is the important development which has intruded on the somewhat visionary yet economically sound policy of freely moving trade, capital, and persons. It is reflected in both the development of international business firms and the vast increase or explosion in research and development efforts and expenditures. Although the two are connected, they may be thought of as twin forces generating tensions in international technology and science which are reflected by the popular jargon of technological gap and brain drain.

The tension of the technological gap stems from the nature of multinational businesses, headquartered in the United States, which seem to possess such superior technology and capital that they are invincible competitors to the national businesses of the developed countries, and appear as powerful octopi threatening the very existence of the underdeveloped countries. Yet the other side of the coin is that these giants are constantly transferring technology to the "threatened" country to close the gap. On the other hand, the brain drain is more directly related to the research explosion in the United States itself. No doubt some engineers and scientists are pirated by foreign subsidiaries of American business. Chiefly, however, scientists and engineers migrate to the United States to find opportunities either in the burgeoning research structure or in dynamic businesses like electronics which are closely associated with or based on research and development activities.

American business has entered the international field comparatively recently. For some time after World War II, when American firms opened branches abroad, they were in Canada or Latin America. With the emergence of a renascent Europe with fast-growing incomes and rising living standards, and faced by the prospects of competition that a large Common Market would entail and possibilities of exclusion by European tariffs, American businesses have gone international. American investments overseas have cumulated year after year, attaining huge proportions. The Department of Commerce estimates that the book value of American-owned affiliates or direct investments abroad for all areas was $49,217 million in 1965 and $64,756 million in 1968. The rate of growth of investments in Western Europe has been particularly heavy. The direct investment of American business in Great Britain, for example, was less than $2.5 billion in 1959, but had jumped to $6.7 billion by 1968.[3]

The extent and influence of these subsidiaries is suggested by the growth of foreign sales of foreign manufacturing affiliates for all areas, which were $20.6 billion in 1959 and $42.4 billion in 1965. For Europe alone, the corresponding sales levels were $7.7 billion and $18.8 billion respectively. While later American government figures are not now published, European estimates place the current totals of all international companies at much higher levels: $90 billion of investment generating $240 billion of annual sales.[4] Proportionately, the American share of sales would be moving close to $200 billion, larger than the gross national products of most countries.

Profits are small in volume, relative to the total of domestic business in the United States, but in the case of individual companies, they may be very substantial. Influence abroad is further magnified by the fact that the foreign-based company is not only controlled by American interests, but the subsidiary itself may act as a holding company for other enterprises within or across national boundaries. American influence in Western Europe's automotive industry, for example, was estimated at 21 percent of the output in the Common Market, and 52 percent of Great Britain's output derived from plants controlled by Ford, General Motors, or Chrysler.[5]

American companies hold that all this is a healthy, competitive state. They argue that their subsidiaries improve the European countries' balance of payments through exports and import substitutes. Americans maintain that they stimulate native industry with the competition they introduce, that they also introduce new technology, new manufacturing methods, and the benefits of the latest American research and development efforts.

The recipients are not so sure. Sir Paul Chambers, head of Britain's Imperial Chemical Industries, described American domination as "a menace to the whole economy of Western Europe." [6] In a speech before the House of Commons, Prime Minister Harold Wilson declared that Europeans do not want investment that brings domination or, in the last resort, subjugation, and he boasted that his government had saved the British computer industry from the predatory actions of American firms.

Naturally, Americans are disturbed by this outspoken resentment as the government's established policy of free movement of capital is being called to account. Government and business committees have been studying the problem, and some modification of the unrestricted policy was considered in the light of these fears which are evoked not so much by the firms themselves, but by their superior efficiency, produced by the new research-based technology. Relief has come in 1968 for a different reason with the imposition of American controls on foreign investment to protect the balance of payments, but the growth of the euro-dollar market offsets its effects.

THE RESEARCH EXPLOSION

Just as the multinational business is the outward manifestation of the new technology, the research and development effort explosion within the United States creates the magnet which draws scientific and technical talent from abroad. In the past decade, the growth in expenditure on R & D efforts has increased enormously. It is estimated that in 1967 the amount invested in research and development is twenty times the amount invested before World War II.

Between 1954 and 1962, the rate of growth was about 9 percent per year. It is concentrated, according to a National Science Foundation survey, in military, space, and atomic energy-related industries. Particularly cited are aerospace and aircraft industries, military electronics, communications, and nuclear energy.[7]

While most of the research is financed by the United States government, and three fifths of all R & D is funded for the defense or space agencies, private companies carry on most of the research of the applied and development type. This creates a nexus with the overseas activities of American business. Often the company with military research contracts sets up new subsidiaries abroad, as the fruits of such R & D efforts give the company an advantage abroad as well as in the United States. The foreign country also is likely to be particularly susceptible to the products of the research or the by-products of some spectacular finding in space or military research, as in the case of the Texas Instruments Company described earlier. Through a subsidiary or affiliate, the foreign company can also exert a reverse flow of research findings. However, the big American company tends to centralize its research in the United States, a policy contributing to foreign resentment, although in most cases large scale research antedated the movement of American business interests overseas. Yet the addition of foreign subsidiaries or affiliates can often add a new production line or new development. The company's interests are broadened and its image of itself sophisticated by its international environment. The tendency of American industrial giants to think of themselves as in some functional activity, such as communications, instead of a specific line, such as electronic equipment, may have been influenced by the international activity. In turn, the company's own image of itself influences its research objectives, the kinds of work done in its laboratories, and the mix of scientists and research men it employs.

The research activity has increasingly come to occupy a central place in the modern American corporation's vision of itself and in its functioning activities. It is no longer the peripheral activity of a few years ago, but a recognized agent of maintenance of company position and growth. This is reflected in the great number of scientific men and engineers who are now on the decision-making levels. It is not too much to say that the research and development function is near the heart of the company, a fountainhead of ideas for its formulation of itself and its planning horizons.

As noted, research activity itself is heavily financed by the federal government, and reflects the social goals of society, as embodied in the allocations of resources to given ends. The character of the research also reflects the American temperament which stresses the applied side—results—even at the risk of drying up inspirations derived from the basic science side. Typical of the constant pull toward applications is the university scholar or researcher who is often drawn away from the theoretical side of science into the government-business-controlled complex of the research explosion.

Similarly, the brain drain draws scientists and engineers of all levels from their own countries to the United States. This is matched by regional brain drains in the U.S. itself, as from the Middle West or South to the East and West coasts. The burgeoning research process itself attracts many, but the whole process, from research through development, testing, and evaluation to the final product utilization, requires increasing numbers of technicians of all kinds. High salaries paid by American corporations are only part of the lure. The chance to be part of the technological revolution, to work with first-rate equipment, and to enjoy greater fringe benefits and more chance of promotion than in his own country also lures the foreign scientist to the United States. Many feel that the military-related research technology in the United States is a new, exciting challenge and opportunity.

American business does not hesitate to seek out talent when needed. It has been estimated that over the 1960 decade as a whole, the demand for engineers, including engineering graduates, other college graduates, and non-degree personnel, is expected to average 72,000 a year, compared with a projected available supply of about 45,000 a year.[8] Particular shortages are generally related to the degree of the research base—as in aircraft, space, and electronics. American companies advertise for engineers in foreign newspapers and send recruiters abroad with excellent results. Immigrant engineers are about 11 percent of the total annual supply of graduate engineers in the United States. One foreign observer notes the "chronic failure" of American universities to supply engineers to meet the demand. "Without foreign scientists and engineers, American technology and economy would not be what they are today." [9]

THE PULL OF THE MAGNET

Chiefly, it is the magnet of the new, military-related, research-based, corporate technology in the United States during the 1960's which has caused the brain drain. Other countries resent both the invasion of American business into their countries and the American drain of talent which the foreign nation's educational system has taken enormous time, effort, and expense to produce.

To understand the problem, one must study the nature of the magnet—this new technological system. Is it new or is it an intensification of trends peculiarly American? Much can be said for the extension theory. From the time of De Tocqueville, the genius of the United States lay in application, in getting quick results. The knack of getting something to work more efficiently instead of more durably, to produce in quantity cheaply without sacrifice of quality, stand out in American industrial history. The ability to manage research, to integrate it into the corporate process, and to move the fruits from theory to the commercial market place is an extension of this basic, historical trait of American ingenuity.

It can be argued that the new technology in the United States, based on research, is but an extension of a system peculiar to the conditions of American society, a matter of American factor endowment. Foremost is the historic scarcity of labor, which forced the development of labor-saving machinery and other such innovations. To this day the American firm endlessly tries to offset the high cost of American labor, giving rise to the bogy of automation. Those industries which have been successful in automating are the competitive industries. Those which are weak, protected, and subsidized usually are those with a costly labor factor.

A second condition in the national endowment, but not unrelated to the first, is the large size of the American national market in a geographically huge nation where goods pass freely over state lines without the tariffs and other fragmentations of petty autonomies so characteristic of the many small nations of the world. Large firms, with large finance, credit, marketing organizations and other corporate paraphernalia, have been made possible by the scale of market conditions. Large-scale production with its well-known economies, internal and external, has resulted chiefly from the advantages of a large national market. One economic historian singles out this factor as the essential ingredient of U.S. growth.[10]

A third element, peculiar to the United States, has been the relative absence of fixed tradition, and a relative openness of society which made the people themselves willing to accept innovations as well as fostering those who enterprise them. There was an absence of feudalism and class heritage. Business enterprise was rooted in the American heritage, and the ideals of hard work, thrift, accumulating, and "getting ahead" in the world were endemic. The entrepreneurial spirit, that much valued element of development, also was important, as exemplified in the Horatio Alger legend, which prized not only hard work and thrift, but undertaking new activities as well.

Further, these American pioneering activities had the benefit of cumulation over a long period, and such tendencies were intensified by the demands of World War II. In one crucial area, the mass education and training of manpower put into practice new ideas of mass training which developed whole new areas such as visual aids, training films, etc., but these innovations were grafted onto a mass system which had been worked out long ago. The land grant college or state university in the Middle West is the lineal predecessor of mass training developments.

Similarly, concern for the consumer is a characteristic of modern, service-oriented marketing methods which seek to create and satisfy consumer demand. This consumerism pampers him with after-sales care, service, and guarantees. It springs from democratic notions of respect for the individual or awareness of his retaliatory power through the vote if he were aroused.

The federal government's role of financial sponsorship of research and development efforts was preceded by setting up industry standards for specifications, procurement procedures, quality controls, etc. The government has long molded conditions favorable to vital private enterprise, establishing the antitrust program, assisting small industry, and regulating the securities market and various industries so as to maintain a high standard and equitable performance. Government sponsorship of R & D activities is framed in such a competitive, standards-setting background.

The foregoing are some of the conditions, peculiar to American society, which are closely allied with the new technology. It is not easy to extricate technology from such factors, yet the new technology itself is so startling that it is almost a thing apart. Many thinking people believe that this new, research-based technology has an identity apart from the institutional framework and that what is happening in the United States is a preview of the future for all mankind. The developed countries will experience the effects of the new technology, forcing changes during the next ten or fifteen years. The underdeveloped countries will feel the effects over a much longer span, providing they make the correct decisions and great efforts to use the new technology.

Some persons even believe that the effects of the new technology will be so drastic as to refashion man. The fantastic development in telecommunications, for example, has given people an immediacy of contact that previously would not have been imagined. The work of the computer, now in its infancy, promises incalculable effects on man's control of his environment. One can hardly visualize the future twenty years hence, but great changes are inevitable. It is even possible that McLuhan's vision of a return to tribal man may not be as farfetched as it seems.[11]

To summarize, the American superiority in military-related, research-based technology stems from conditions which are largely peculiar to the United States, and the new technology is concentrated in a sector of industries heavily supported by government policy. Yet the new technology has an identity of its own, generating strikingly different conditions. Within the United States itself the new technology is causing profound changes within the corporate structure, the universities, the cities, minorities, transportation, and other sectors. Internationally, the new technologies are also causing disturbances, due to their startling character and effects. The United States, without intention, has become an object of envy and resentment as well as a magnet attracting forward-looking people from all parts of the world. Moreover, rightly or wrongly, it is often thought to be a model for mankind's technological future. Circumstances have forced the United States to occupy this sometimes uncomfortable role, but it is a fact of twentieth-century life from which a world technological system is emerging.

NOTES TO CHAPTER 4

1. Department of State, *The Battle Act Report, 1964,* U.S. Department of State Publication 7736 (January, 1965), pp. 29–33. However, some relaxation of controls is evident in act's renewal in late 1969—*New York Times,* January 11, 1970, p. 6.

2. U.S. Congress, House of Representatives, Committee on the Judiciary, *Immigration and Nationality Act with Amendments* (6th ed. Washington, D.C.: U.S. Government Printing Office, 1969), Titles I, II.

3. U.S. Department of Commerce, *Survey of Current Business,* XLVI (September, 1966), 34–35; *Ibid.* (October, 1969), 23–28.

4. U.S. Department of Commerce, *Survey of Current Business,* XLVI (November, 1966), 8; *New York Times,* January 16, 1970, p. 51.

5. *Business Week,* January 27, 1968, p. 143.

6. *Times* (London), January 20, 1967, p. 15.

7. National Science Foundation, *Federal Funds for Research, Development and Other Scientific Activities,* Surveys of Science Resources Series, NSF 66–25, July, 1966, p. 3; *Reviews of Data on Science Resources,* NSF 69–12, February, 1969, p. 4.

8. National Science Foundation, *Scientists, Engineers, and Technicians in the 1960's,* prepared for the National Science Foundation by the U.S. Department of Labor, NSF 63–64, 1963, p. 2.

9. Alessandro Silj, "Should Europe Recall Its Scientists?" *European Community,* No. 127 (September, 1969), p. 10.

10. Douglass C. North, *Economic Growth of the United States* (Englewood Cliffs, N. J.: Prentice-Hall, 1961), p. 166.

11. McLuhan cited in Chapter 2, footnote 11.

5

DEVELOPED COUNTRIES: PRIDE AND PREJUDICE

For some time the developed countries of Western Europe have indicated that they feel they are on the losing side of the so-called technology gap and brain drain. The technology gap has been discussed in North Atlantic Treaty Organization (NATO) meetings and by the European Parliament, governing body and part of the European Common Market, and it is a subject of study by the Organization for Economic Cooperation and Development. Amintore Fanfani, then Minister of Foreign Affairs for Italy, proposed a Marshall Plan for closing the gap. Harold Wilson, Prime Minister of Great Britain, associated Britain's effort to enter the European Common Market with a proposal for a European technological community to pool research findings and applications for a huge, integrated European market. The United States government, reacting to European concern over the alleged American technological empire, has been debating the matter in government committees, at high and low level, and in the forum of the press and periodical articles. Indeed, a European official told the author that Americans seem "overconcerned."

What is behind the European contentions? What is the American image in Europe? What is the reality behind the image? What are the choices? Basically the Europeans and the British seem to feel that since the United States spends an estimated four times or more each year on research and development activities than does Western Europe, the advanced European countries eventually will be reduced to the position of satellites of the United States, economically and technologically.

According to Prime Minister Wilson, there is being established:[1]

> . . . an industrial helotry under which we in Europe produce only the conventional apparatus of modern economy while becoming increasingly dependent on American business for sophisticated apparatus that will call the industrial tune in the nineteen seventies and eighties.

Western European countries, long a source of technological innovations, now are upset to find that much of their technology, particularly in space-age industries, is dated and relatively infertile as far as new technological developments go. They particularly fear American domination (even subjugation) in industry which occurs as American companies acquire holdings in European companies which can no longer compete. Europeans believe that the loss of their scientists and engineers to American companies is connected with superior American research and technology and particularly to the policies of companies which center their research activities in the United States.

No doubt there is some reality behind the European fears, but there is also a positive side. The system which Western European countries share with the United States is basically a market system in which the interplay of forces, shaded by judicious government polices, decides matters. Sometimes market decisions go against particular companies; often there has been failure to compensate with judicious intervention or there is a lack of research allocation, technical practice, or application. There is a gap in opportunities in many countries for young engineers. All of this points up the need for precision in defining such gaps analytically and statistically, and such study is underway, but these gaps are not necessarily undesirable. Assuming that the goal is to improve a local technological capability, the gaps may provide the prodding a country needs to advance technologically. Instead of being a threat, the gaps indicate the lack of attention to internal needs on the part of many European countries. The existence of gaps may be a blessing if they stimulate advanced countries to improve their techno-economic conditions. Fear of change is understandable, but it should not reinforce a complacent status quo attitude. Technology has an independence and a determinism to which all mankind must adjust sooner or later. The sooner a nation realizes this fact the smoother its technological path will be. Contrariwise, those nations which cling to national pride and prejudice are likely to be hurt. The case of Great Britain offers both forward-looking and change-resisting elements which are representative of most developed countries under the impact of the dynamic, new technology.

THE UNITED KINGDOM

The United Kingdom pioneered the industrial revolution. The steam pump was invented because more effective pumps were required to remove water from Cornish tin mines. Soon after, Thomas Newcomen invented the steam engine, which James Watt rendered technically effective. The firm of Bolton and Watt, after many commercial struggles, brought business know-how to bear, making the steam engine a commercial success. During most of the nineteenth century, British engineers, scientists, technicians, and businessmen had the creative drive, ingenuity, and business acumen that made the British supreme both technically and commercially. It was a British century, technologically and politically. No doubt in those days other countries must have felt the lag as a technological gap, but the term had not yet been invented.

Yet other countries did something about the situation. After the American Civil War and the unification of Germany in 1870, the United States and Germany launched a technical development drive which eventually enabled them to outdistance Great Britain. At the same time, the very success of the British in technology generated a complacency which resulted in failure to make adjustments necessary to meet the competition. It is significant that a British economic historian has assembled evidence to show that British corporate enterprise lost its dynamism early.[2] This source shows that from the turn of the century voices were heard complaining of the lack of vitality in the techno-economic sphere in very much the same terms as are heard in modern Britain. The roots of British difficulties go back a long way to decisions at some point which traded off continued rapid technological progress for stability and conservatism in society.

Modern examples of this trade-off phenomenon may also be documented from a plethora of contemporary commentaries by British and foreign analysts. Anyone who reads current British newspapers and periodicals can see that the British are increasingly aware of their technological dilemma. Three examples will be given to show how the British opt for current comfort at the price of technological advance: first, the Rootes industry case; second, the personal option of a distinguished British scientist to emigrate; and third, the lack of encouragement of the British inventor Barnes Wallis. Taken together, these examples sketch a kind of discouragement of technological vitality from which Britain seems to be suffering.

Rootes Motors, one of the British big five in the automotive industry, was losing money, and the Chrysler Corporation offered to assume majority control, raising its share holding from 45 to 65 percent. After much discussion the British government reluctantly approved the arrangement while retaining some surveillance through the government-owned Industrial Reorganization Corporation (IRC) by buying a share and appointing a director to the Rootes board. This decision meant that more than half of the automobile industry in Britain was controlled by the big three automobile companies in the United States.[3] Before agreeing to the Chrysler take-over, the British government consulted with leaders of the main British-owned auto firms to find some means of keeping Rootes a British-controlled company. Even nationalization was considered. However, no alternative plan could be worked out, and the government was forced to accept Chrysler's offer, but only after Chrysler assured the government that the majority of the directors would be British. The decision aroused a barrage of criticism from the British press and people. In defense of the government, the comment was made that the Rootes deal with Chrysler was made because Chrysler might otherwise have expanded the Simca plant in France, where Chrysler controls 76 percent of the stock. Yet the British are distressed to know that an American company controls 52 percent of the stock in a national industry of prime consumer importance. But this condition is only a little more extreme than in other advanced countries. A third of West Germany's motor car output is also American-controlled. To the British and the Europeans the Rootes case was a symbol of what is happening everywhere.

The case shows a grudging acceptance of the technological and managerial benefits of the vigorous United States competition in order to invigorate this declining company. Implicit in the decision is recognition of the benefits accruing from American manufacturing methods and research and development. Exports will be benefited; indeed, some of them may go to the United States market itself, in competition with the parent company. The Simca alternative is recognized realistically. But the question still remains: If preservation of national control was so important to the British, why was more of an effort not undertaken to develop an effective alternative?

The other side of the coin from the technological invasion is the exodus of scientific and technical personnel. Examples can be found almost daily. Thus, a distinguished medical scientist, Dr. Frank Kingsley Saunders, director of the Medical Council Virus Laboratories at Carshalton, near London, announced that he will leave for a laboratory built especially for him at Sloan-Kettering Research Institute in New York. In England he has a staff of thirty; in New York he expects to have two or three times that number. His salary will be increased from the $12,600 a year he received in England to $27,000. Stating his reasons for leaving, he said:[4]

> My reason for leaving is simply opportunity. Money by itself should not be the main motive for any scientist to go anywhere. The important thing is that I will be able to have young graduates working with me in America.

Since colonial days British scientists and technicians have emigrated to America, but it is only recently that the trickle of emigrants has become substantial and held to be injurious. Statistics are not available in detail, but one widely quoted report shows that in 1961 the annual rate of scientists and engineers with doctorates leaving Britain for good had reached 17 percent, of which 7 percent went to the United States. These men were of high quality.[5]

People with lesser qualifications also leave in substantial numbers. A British survey estimates that they include 8 percent of all engineers and scientists taking their first degree. While small in comparison with a total of 30,000 such personnel, it is an uncomfortable percentage of graduates and first-job holders.[6] Although the total proportion of men with some years of experience who emigrate is not known, the number in certain industries is believed to be high. The Society of British Aerospace Companies reports that 1,300 qualified specialists left the aircraft industry in 1966, with more than 800 going to the United States or to American-owned companies.[7] These key men were in air frames, missiles, jet engines, space, electronics, and equipment fields. Their reasons for leaving were said to be lack of certainty of opportunity in future industrial research, development, and production, as well as wage differentials and other lures. One conclusion which was reasonably drawn was that the British government had failed to support its aerospace industry's development on a scale and time horizon to meet American competition.

In this light it is instructive to examine the case of Barnes Wallis, outstanding British inventor.[8] Ten years ago Wallis invented the variable wing supersonic airplane and sought to develop it under government research contracts. This wing is based on the same principle with which the Boeing Aircraft Company won the contract for the supersonic transport plane on January 1, 1967. When Wallis invented the variable wing, he was an established inventor who already had designed such successful aircraft as the Wellesley and the Wellington. Among his other inventions was the bouncing, dam-bursting bomb. However, Wallis failed to get the backing of the British government for the variable wing, supersonic transport plane, and his plan for 12,000–17,000 mph jets (London to Australia in one and a half hours) seems also to have been shelved as too costly for development. Taking this decision in conjunction with the exodus of aerospace personnel, it is difficult not to conclude that the decision-making elites in England lack the imagination necessary to adapt to the modern, technological world and its fast competition.

What causes this obviously able nation to lag behind? Britain appears to discount technological progress in favor of stability and conservatism. "Too comfortable" is the pronouncement of one analyst and his research team.[9] The elite, government officials, universities, labor groups, and the people as a whole seem to be endowed with an unshakable solidity and reluctance to accept change easily. While this conservatism can be a source of strength, it is also a handicap in a world demanding continuous technological change.

The elite classes who govern the country and manage England's enterprises have well-defined ideas of what is important to them, which set the pattern for the rest of the country. If the elite considers holidays as sacred, as dedicated to rest without work, the masses will follow. The conservatism of the elite is born of centuries-old security. When outside individuals or their children are admitted to the Establishment, they soon assume the values of the elite. Where a man has gone to school, or his pronunciation, may be as important as his ability to work. Deference is rated above productivity or critical analysis. Although such social restrictions exist elsewhere in the world, they also show why the British and other advanced nations in Europe have seemed unwilling or unable to learn the lessons of American technological progress.

The government officials in key positions are drawn from the elite class, trained in "good schools" and holding "good degrees," which mean they have been educated in the arts and social sciences, particularly economics. They are cost-conscious and tend to emphasize economic utilization of existing resources. They often are oriented to finance rather than to production or management problems. Lineal descendants of the men who directed the destinies of the now largely liquidated empire, they appear cautious and conservative. British civil servants are often hailed as brilliant, and no doubt are as individuals, but collectively they do not seem to think imaginatively, at least on matters of science and technology. These are the men who advised caution on Barnes Wallis' proposals.

Although the management of British industry doubtless includes men who are the equals of the best in the United States, a recent British study suggests that these are the exception, not the rule.[10] This study points out that management practices must be improved and modernized if British industrial efficiency is desired. In the United States, the business schools of the better universities have been giving management training courses since before World War I, and such courses have been taken for granted since the 1930's. These management training courses encourage men to make decisions, stress ability and results, and discount deference to seniority and status. In Britain, almost the opposite is true. To serve one's country by entering the civil service carries more prestige than a management position in private industry, and even private industry itself emphasizes preserving the Establishment. Thus, recent attempts by some productivity-conscious leaders in British industry to recruit youth trained in American business methods have met with little success. It is rumored that Lord Nuffield's attempt to found a business school at Oxford University was rebuffed, yet such shortsightedness can be costly in an age when managerial practice is becoming increasingly demanding and sophisticated. However, some British managements have begun to be somewhat more flexible and open-minded toward better practice whatever its origin. Thus, Hitachi of Japan and Wellman Machines, a British crane maker, recently signed a technical assistance contract on how to build special cranes for the steel industry. British steel executives saw the crane on a tour of Japanese steel mills. The resulting contract has been hailed as the first time the British have accepted superior Japanese technology.[11]

Until recently, Britain had a narrow-based educational system which restricted opportunities for the majority. A scholastic test given to students at the age of eleven determined whether persons would go to grammar or college preparatory schools, or leave school at age 15. This test restricted opportunity for "late bloomers." As a result of the publication of the Robbins Report,[12] recommending the expansion of the higher education system, upgrading of technical schools, and the creation of more universities beyond the elitist few, the system is now changing. However, the expanded system will continue to favor arts and sciences against technology.

The traditional, nice distinctions between the gentry and the working classes have extended to science, which is preferred to engineering. "Pure science" has been much admired; engineering or technology has been considered a lower level activity. There is lack of public understanding of the continuity of work from concept to finished product involved in modern research and development efforts, especially in industry. In 1966, the manpower survey began to treat engineers and technologists as equals in an official document that recognized also the importance of the technical supporting staff.[13] However, the recognition comes late.

In reviewing the science policy in Great Britain, the Council for Science Policy in its 1966 report recognized the underlying rigidity when it called for:[14]

> . . . a greater readiness to redeploy existing resources and manpower upon new projects, and generally a much greater flexibility of employment and acceptance of change . . . than has always been achieved in the postwar phase of economic growth.

At the outset, this report recognizes the autonomy and authority of the national science policy research councils (established in 1918) in setting priorities without considering prior national needs. This sets the tone of the report, which also stresses the theoretical side of research. Although the report discusses a "sophistication factor," citing the change from optical to electron microscopes and the adoption of complex electronic data-handling equipment, it ignores the roles of contract development or mission-oriented, industrial research which is so important to maintaining economic viability, closing technology gaps, and alleviating the emigration of trained personnel. This report is an interim one, and the commission may develop "sophistication" about something other than equipment. However, the initial report reflects the British concern with economic utilization of existing resources instead of seeking ways to build larger ones. To think of science in terms unrelated to final products and industrial growth is to isolate science unrealistically.

If a top level report written after much study of American practice still does not advocate or see the need for the essential research cycle from concept to completion, it becomes clearer why British scientists and engineers are so easily lured away. Modern technology is often the end product of scientific activity. Its creative effects on the economy in accelerating national income growth and permitting still larger allocation of resources are of the essence of the interacting process. The British *Report on Science Policy* simply does not come to grips with the fundamentals of an advanced economy.

WESTERN EUROPE

That other advanced European countries have experienced technological difficulties similar in principle though not in detail to those of Great Britain does not require elaboration. They have responded more or less progressively to the challenge according to their traditions. All European nations are alarmed by the dual pressures of the invading American technology and the exodus of many European scientists and technicians. The former is the more sensational, but the latter is an unhealed wound.

Ten years ago, before research technology was perceived to be the root of the American lead in key industrial sectors, the European Common Market was established to meet the need of a large market. The example of the United States showed Europeans that a large-scale market makes possible external and internal economies which reduce costs and increase income. The success of the Common Market is evidence that the plans have worked. In July 1968, when the internal tariff barriers were removed, the free movement of trade and persons became complete. The increase in the external tariff has been ameliorated by the Kennedy Round of tariff reductions. However, President Charles De Gaulle of France spearheaded a revival of nationalism among European countries, and any supranational government is unlikely in the foreseeable future.

This political weakness affects the European response to the technological challenge. In the case of the United States, the technology is based not only on the large-scale market, but also on a substitute for the market—large-scale, central government procurement of military, space, or nuclear research and development—which has commercial application or spillover effects. The European Common Market is considering the matching of this expenditure by pooling the efforts of various member countries, but the nationalist revival is a handicap. Robert Marjolin, for example, has drafted an unpublished report that urges European governments to combine their orders for advanced products to enlarge the markets of the Common Market members.[15] The report also advocated creation of national government corporations to manufacture technical products, such as computers, and recommended the establishment of an intercountry educational center of technology modeled after the Massachusetts Institute of Technology.

The proposals drafted originally in this document are reported to have been watered down for fear of infringing on the sovereignty of member states, especially France. The final version merely stresses the need for national research contracts, the use of national but coordinated tax and incentive policies to spur development, and the possible development of joint and coordinated projects. The report also proposes greater interchange with the United States and possible cooperative ventures. However, the report warns that such cooperative activity may even widen the existing technology gap. This warning reflects the ambivalent attitude prevalent in Europe which wants American technology but fears American domination.

The attitude of Common Market members cited above reveals that it may not be politically possible for them to match the American conditions to develop their own technology. Still, the development of the Market itself has set a precedent for international cooperation which may also be applied to developing technology despite local nationalism. European nations already have cooperated in such specific fields as nuclear energy, space, the Anglo-French supersonic plane and the Franco-German A-300B jumbo jet. Europeans may cooperate more closely, and if pressures build up, it is possible that a new European type of research-based technology may emerge.

The British proposals to enter the European Common Market have been related to technological cooperation and vitality on a large, long-run scale. The British believe their strength lies in basic research, which, joined to the efforts of the European Common Market countries, can offset effectively the American lead in technology. However, there is the contrary argument that there is little affinity between British and Continental systems of thought and practice and that therefore the United States may be a better partner for Britain in technological activities, due to the basic Anglo-American heritage of ideas, traditions, and language.

The European nations, especially France, recognize this Anglo-American affinity and regard it as a good reason to deny Britain's proposed entry into the Common Market. A kind of Trojan horse situation is visualized which would open the door to still further control by United States-based international companies, relegating the nations of Europe to less important roles in developing new technology. It may well be that Britain's efforts to join the Market will founder on this point.

All of the European nations fear the domination of any supranational union, and there is genuine danger in giving too much authority to bureaucrats who are further removed than a national government from the controls of an electorate. But most of the distaste for a supranational union stems from the reluctance of each country to accept encroachments on its national sovereignty. This reluctance is reinforced by European uncertainty about American corporations. However, American business may reach the point of diminishing returns in making investments in Europe. Therefore, the need of grand methods to counter the American business invasion may not be necessary. For one, the controls over further dollar investments directly from the U.S. have cut the flow substantially. Moreover, since 1959 Europe has had a continuous monetary inflation, and American business may find there are fewer chances to make high profits, so that a slackening of American investments in Europe, even in the absence of controls, is quite possible. Likewise, an American recession could slow down or halt the current brain drain from Britain and Europe as the American demand for more scientists and engineers would lessen. American companies which already have experience in Europe may decide that European labor has a lower productivity that more than offsets their lower wages. Lastly, by 1969 the profit rate for American companies manufacturing in Europe had fallen to levels prevailing in the United States.[16]

But even if these pressures are lifted, the situational logic for the European nations is to continue their plans to integrate and develop the Common Market. Whether Britain and her partners in the European Free Trade Association (EFTA) join the original six or move toward an Atlantic community with the United States is not foreseeable. However, it is clear that the advanced countries, in their own interests, must take new social and economic measures to meet the challenge of the new American technology. Pride and prejudice, inherited from outmoded social systems, must be jettisoned to accommodate to technical progress and to keep the economies of Britain and the European countries viable.

ECONOMIC CHOICES FOR ADVANCED NATIONS

The advanced countries of the world face various choices regarding science and technology. One may question the wisdom of Europeans who insist on making their own research and development efforts in the prestigious areas of computers, aerospace, and nuclear energy. It is possible that European nations may concentrate on research in those areas which are not yet being exploited by the military-space complex. This technology may be licensed from the United States in many of the latter cases. Japan, for instance, already has shown the way in her shipbuilding, electronics, and other industries by the maximum borrowing of technology and a minimum of supplemental research. In such a case the borrowing country may concentrate its own research efforts in other areas which may have potentially higher payoffs.

Conventional or traditional industries now have little research and development resources allocated to them. For example, the construction industry certainly stands in need of innovations which would make housing for the masses more economical and attractive. Conventional industries also provide a much larger share of foreign exchange earnings for the balance of payments than do the glamour industries. A country seeking to strengthen its foreign balance would find that the yield of incremental investment in conventional industries is much higher than the profit from duplicating typical American research efforts.

The British, who rank second to the United States in their expenditure on research and development efforts relative to their gross national product, are concerned about the payoff of this investment.[17] The British now spend 2.2 percent of their gross national product on research, compared to 3.1 percent in the United States, 1.5 percent in France, and 1.3 percent in West Germany. One British observer believes the conventional industries could provide four times as much in exports for Britain compared to the glamour industries, yet would cost only half as much in research and development expenses.[18] He calculates that a 10 percent increase in automobile sales would have much more productive effects than the same increase in products of a small, science-based industry. This analysis establishes a valid principle, namely, that the returns of emphasizing research that will increase national income and improve the balance of payments make possible a larger commitment of resources to research at a later period. Discussions of the need to reallocate existing sources for research and development purposes often overlook the principle of investment for growth in the near term.

A second choice the advanced countries have lies in the proportion of resources they can allocate to research and development.

Both absolutely and relative to the size of their national products, the amount of resources allocated to research in all advanced countries is less than in the United States. This is a significant fact. There are other factors, but generally an effective effort in research is a function of the resources committed to it. To opt for greater research effort means to let something else, for example, high-speed rail transport or highway building, go by the board if necessary. If a country wants results, it must pay for them. If it wants to retain scientists and engineers, it must pay higher salaries, give more status and respect to budding, younger men, promote more quickly, and alleviate the heavy hand of bureaucratic elites. More investment in the education of engineers, technicians, and managers is another requisite for results.

The concern immediately involves a third area of choice—the amount to allocate as between applied areas and basic research. If the former is emphasized, there is a good chance of dividends which will offset the investments. More theoretical research is not likely to have immediate effects, though there are exceptions. It has been estimated that most of the science in use today was known fifty years ago. If an advanced country wants to close technology gaps, its choice should favor the applied side of the spectrum. Basic research can be increased later. In this case the example of Japan, which has borrowed heavily and directed much of its own research and development to improving its practice, provides a lesson for the advanced countries.

There is a place for basic scientific research, but the chief gaps between the United States and the other advanced countries are due to the fact that the other nations lag in industrial management, engineering, finance, marketing techniques, and the willingness to risk new ventures. The British and the Europeans did much of the basic research in nuclear energy, jet propulsion, electronics, and in many other fields. However, as the case of Barnes Wallis in England demonstrates, they failed to capitalize on the inventions and discoveries which they made.

Other choices face all countries in deciding the proportion of research resources to be allocated between many sets of polarities. Some of the choices are: military versus civilian, natural sciences versus biological sciences, hard science versus social science, concentration in large centers versus dispersed geographical locations, and contract versus in-house government research. Probably none are as important as the decision for more international organization and cooperative large-scale effort. Some research by nature requires such a large effort that it may be beyond the capability of a single country. The space effort is the well-known example here. The French effort, the only European national effort of consequence, duplicates much American and Soviet research. A decision to opt for international cooperation may be essential, but the difficulties and resistances likely to be encountered loom very large.

NOTES TO CHAPTER 5

1. *New York Times* (International ed.), March 14, 1967, p. 1.

2. D. H. Aldcroft, "The Entrepreneur and the British Economy, 1870–1914," *Economic History Review,* Second Series, XVII (August, 1964), 113–34.

3. *Business Week,* January 27, 1968, p. 143.

4. *The Evening Star* (Washington, D.C.), December 19, 1966, p. 4.

5. Sir Gordon Sutherland, "The Brain Drain," *Political Quarterly,* Vol. 38, No. 1 (January–March, 1967), p. 52.

6. *Ibid.,* 54.

7. *The Times* (London), February 1, 1967, pp. 12–13.

8. *New Scientist,* January 26, 1967, p. 192–93; *Asahi* (English ed.), February 21, 1967, p. 7.

9. "Herman Kahn's Thinkable Future," *Business Week,* March 11, 1967, p. 115. Herman Kahn and Anthony J. Wiener, *The Year 2000* (New York: Macmillan, 1967).

10. Anthony Gater and others, *Attitudes in British Management, A P.E.P. Report* (London: Penguin Books, 1966).

11. *Japan Times,* March 2, 1967, p. 8. Another prominent British businessman declares that the "copycat image of Japan" is now dead. *Asahi* (English ed.), March 8, 1967, p. 7.

12. Great Britain, Committee on Higher Education, *Report of the Committee Appointed by the Prime Minister under the Chairmanship of Lord Robbins, 1961–63* (London: Her Majesty's Stationery Office, 1963) Cmd. 2154.

13. For comments by a British scientist, see Maurice Goldsmith, "The Autonomy of Science: Some Thoughts for Discussion," *Political Quarterly,* Vol. 38, No. 1 (January–March, 1967), pp. 81–89. The preference for science over engineering is also found in U.S. funding practices, but to a lesser degree.

14. Great Britain, Council for Scientific Policy, *Report on Science Policy* (London: Her Majesty's Stationery Office, 1966), Cmd. 3007–8.

15. *New York Herald Tribune* (International ed.), March 18–19, 1967, p. 2.

16. F. Newton Parks, "Survival of the European Headquarters," *Harvard Business Review,* Vol. 47, (March, 1969), p. 79. However, in spite of seeming

easing, Raymond Vernon thinks the pressure still warrants some kind of international controls, *New York Times,* January 14, 1970, pp. 61, 67.

17. Council for Science Policy, *op. cit.*

18. J. D. Bernal, "Public Policy and Science," *Political Quarterly,* Vol. 38, No. 1 (January–March, 1967), p. 18.

6
THE DEVELOPING COUNTRIES

The euphemism "developing countries," which is applied for face-saving reasons to the two thirds of the world which are poor and underdeveloped, is beginning to sound ironical. While the Orwellian view of these countries as an endless reserve of cheap labor for the developed world is not yet a fact, the twin scourges of increasing population and inadequate food supply (even with the "green revolution") have checked the progress of these countries toward the modest goals of the "development decade." Yet coincident with this stagnation or near stagnation, events of immense technological importance are taking place in the United States and the developed countries. Although the technological gap between rich countries and poor countries is vast, it is likely that new research-based technology offers the best chance for the advancement of the poor countries. Why, then, isn't the gap closing instead of widening? It would seem to be common sense itself to transfer the advanced world's technology as quickly as possible.

Hubert Humphrey underlined the importance of the general subject in his remarks to the *Symposium on Technology and World Trade*.[1] He said at this conference on November 16, 1966, "Long after any North Atlantic technology gap is closed, it will be the business of the Atlantic Nations to try to close the far more dangerous rich-poor gap. We in the rich nations must begin taking more active steps now to help the poorer nations build their economies, create broader markets, and develop their own technologies."

THE PLIGHT AND OPPORTUNITY OF THE UNDERDEVELOPED WORLD

It is hard to identify even the significant factors which in so many different countries are obstacles to importing or obtaining advanced technology. In part, the problem is similar to that of the developed countries: their comfortable elite class fears disrupting change which might strip them of their privileges. The developing countries also lack at least three centuries of historic evolution which marked the rise of the West. The institutional changes which occurred in Western countries—the emergence from feudalism, the establishment of strong national states, the rise of the merchant and business class, the development of rationalistic science and technology—all these and many more advantages have little or no counterpart in the underdeveloped, poor countries. The absence of these firm foundations do not preclude effective development, but it is still a huge handicap. It is significant that the one non-Western country which has developed quickly and effectively—Japan—underwent the same type of institutional change as Western Europe.[2]

Moreover, the developing countries are burdened by ever-increasing populations. In Western countries, improvements in medicine and health care matched economic improvements which allowed the saving of capital and generated higher incomes. In the modern poor countries, the rapidly declining death rate has created the population explosion, while agricultural improvements and increased food production are still inadequate to meet the needs of increasing hordes of the hungry. Undernourished people will spend more on food as their incomes rise, but larger food supplies are not forthcoming to meet the rise in population.

These countries are rent by divisive troubles; caste and class, racial and ethnic divisions, tribal and linguistic differences impose gigantic, longstanding barriers to the integration and stability so necessary to the growth and development of a country. Political differences also are endemic, and it is usual for the government of a typical country to be torn between the need to assert its new, proud nationalism and its need for all kinds of help from the economically vital West. Such a country, proud of its new independence, usually dreads the outside help it needs, in fear that it will entail so-called economic colonialism in its wake. Bureaucratic socialisms, manned by inept and corrupt civil servants, thwart improvement efforts in favor of vast schemes for creating government-owned industries which are long in gestation and slow of operation when started. In short, the poor countries are trapped in a morass of interlocking social, economic, and political troubles.

The saving factor for poor countries could be modern technology and its ramifications. The new technology can help such countries leapfrog historical evolution, and introduce them to new, creative, and productive methods which will revitalize their economies. This technological force is already being felt in many ways. For one, every country today, no matter how backward, has some sort of airline service to its major city or cities, and the airline frequently is based on the country's air force, which is in turn often a product of the United States military assistance program. Modern communications systems also have penetrated to remote areas. However, for a great range of technology, the potentialities are greater than the actualities.

Thus, the population-food crisis potentially can be solved with the aid of technology. Birth control devices of a mechanical or chemical type have been developed commercially, and are capable of ending the population explosion and regulating the number of births. There is also a refreshing realism in the new willingness to accept the principle of birth control, as witness India, which numbers birth control among its chief objectives. Food production has been increased with the proper "green revolution" techniques including adequate fertilizers, irrigation, better seeds, and use of modern machinery. In this area too, the governing elite is showing more realism, but the race with famine is very close.

The technology to deal effectively with the situation often lies with the big international corporations, but obstacles to transfer loom large. For example, large American fertilizer companies, prodded by the United States government, have been willing to go into India in spite of such disadvantages as small scale plants and other disincentives. The Indian government has been very reluctant, however, to come to terms with these interests in spite of its obvious needs. As in the developed countries, there is a fear of being overwhelmed by the foreign interests. In other areas essential to the development of the poor countries, the chances of improving the country by modern American technology are hindered by suspicious nationalists who have indoctrinated the masses with stale socialist slogans calling for redistribution of poverty. Yet the opportunity for the modern developing countries lies precisely in an efficient transfer of technology which must somehow be effected in a reasonable period in spite of obstacles.

EDUCATION, FOREIGN TRAINING, AND BRAIN DRAIN

One of the ways to introduce modern technology into a developing country is to import technicians from developed countries who can fill immediate needs for operating technicians in critical jobs or who can help set up educational and training facilities for the native personnel. Although this simple, direct route is often employed, it suffers from a number of drawbacks. First, operating technicians from the developed countries are expensive; they must be paid more than the going rate they earn at home to induce them to work in a developing country with inadequate facilities and even hardships. Second, technicians often will not stay long enough in the developing country to do the job adequately. Third, the government of the developing country usually is eager to replace the imported technicians with local people as soon as possible, and frequently the replacements are inadequately trained or poorly motivated. Fourth, the imported technician does not operate in a vacuum but needs a complex of capital equipment and supporting technicians to cope with the difficult problems he is supposed to solve. Fifth, the technician must also be a teacher, training the native personnel, and his students and trainees must have jobs waiting for them in integrated, modern plants or institutions which adequately utilize them. For these and other reasons, while imported technicians have a role in developing poor countries, they have limitations.

The converse of the technician going to the country is the student coming forth to study the best techniques abroad. In theory, and over the long run, this may be the best way for a country to learn modern methods, because historically it has always been the educated young men who sought out the best practices in developed countries and then introduced such practices to their own country. The samurai youth of Meiji Japan, for instance, studied in other countries to learn the best methods later used to modernize Japan. Similarly, before World War I, young American physicians gained professional prestige by postgraduate study in Germany's outstanding schools of medicine.

Nowadays, foreign students are easily assimilated into the educational system of the developed countries. The expense of the student's education abroad is negligible to his own government as the students often receive donor country scholarships, work, or otherwise pay their own way. The cost to the donor country is also small, though large numbers of foreign students may cause protests and an increase in their university fees, as happened in 1967 in Great Britain.

In theory, the student from the developing country who has been educated abroad is supposed to return to his own country to spread the benefits of his modern education and training and help to introduce modern ideas and practices. However, in practice the situation often is not so ideal. The educated young man upon returning to his own country may be overwhelmed by the circumstances of old, ingrained ways. His superiors are not eager or willing to see their methods superseded. The young graduate lacks the equipment and supporting institutions which are necessary to help him employ his new knowledge and skills. Thus, a biologist returned to India to head a laboratory, but he could get no equipment for experiments, and he even lacked funds for the barest essentials of his work. It is plain why such a scientist and others like him frequently return to work in the developed countries where they have been trained and where jobs often await them.

Furthermore, in many fields the graduate technician or scientist returns to his native country with textbook technical knowledge, but he lacks the know-how that is acquired only after years of intimate practice in industrial firms. An electronics engineer can hardly set up a plant or open a consulting business without years of solid experience. Even then his chances of creative work would be slender unless he were backed by an established engineering firm.

Under these circumstances it is understandable why the young, ambitious scientist, technician, or engineer decides to emigrate from the underdeveloped country to the advanced country. In the advanced country he finds high salaries, good equipment, and opportunity to develop his professional potentialities. In his own underdeveloped country, he may be faced with actual unemployment, or at best he takes a position at a comparatively low salary, and is forced to work under conditions which his foreign training has taught him to regard as primitive. As he learns of new developments in his field which take place so rapidly abroad, he suffers intense frustration from the knowledge that he is falling behind. His loyalties to his native country usually will not weigh heavily enough if he has a chance to return to a developed country.

Moreover, these facts are even well known to foreign students who are studying in an advanced country. Knowing of the shortcomings of returning home and the rich opportunities for staying in the advanced country, the students find it easy to rationalize that they need more on-the-job training (which is usually true) and decide to delay their return home. The high salary earned in the developed country makes it possible for the expatriate to bring his relatives to the developed country. In this way he can ease his conscience and convince himself that he is doing something to help his poor country, and the problem is put off until the next generation.

The big developed nation is glad to keep the foreign-born specialist if it needs him. It is all in keeping with the Western ideas of individual choice of the free mobility of the worker with his skills. The developed country's government cites its national ideals of nondiscrimination and equal opportunity. As long as there are job vacancies due to full employment policies and technological advances, the well-qualified immigrant will find a job, and will be welcomed into the emerging group of elite professional technicians, and the economy of the advanced country will be enriched by his contributions. Thus, the rich, technologically advanced nation grows richer; the poor, underdeveloped country loses its best asset —its ambitious, highly educated young. The mechanism of spiraling cumulative causation described accurately by Gunnar Myrdal fits this condition.[3]

LANDS LOW ON THE TOTEM POLE

But the serious part of the problem is only beginning to be clear. In a world technological circuit, the sequence which seems to be happening is that the magnet of the United States is pulling people from developed countries like Great Britain and Germany. Replacements for those leaving home come from underdeveloped countries like India and Pakistan which can least afford to lose their services. Of course, the United States itself operates as a magnet directly on Indians trained in the United States, but in either case the lack of opportunities in the poor country contribute to the building of a still larger technical gap through the loss of the services of the brain-drained scientists and engineers from the poor countries. The frontispiece cartoon catches the essence of what is happening.

The remedy, as in the case of the developed countries, is to provide opportunities for these people, but this is not easily accomplished. In the case of scientists, there is little likelihood that centers of the kind necessary to provide, say, a modern physicist with competitive salaries and suitable equipment could ever be financed on any comparable scale. Indeed, the expenditure would be a gross misallocation of resources in the face of the more basic needs of these poor countries. This is not to say that there is no role for basic science in the developing country. Some few will always be needed to act as top scientific advisors guiding the development of the country's educational system in a complex world. They will also have a backstopping role in guiding technical decisions.[4] Such men will also provide inspiration for the youth and see that the educational system is well structured. But obviously the need is for scientists who will work in applied contexts, for engineers and operating technicians.

But even for the applied types, the opportunities are still likely to be very limited unless there is a distinct policy shift on the part of the elites who govern these countries. Like their counterparts in the developed world, they must be prepared to accept changes conducive to bringing in modern technology. Generally speaking, this means coming to terms with the vitality of modern private industry from the developed countries. This reconciliation may not be such an effort or mean such a big accommodation as is believed. There are ways of devising protective screens in the acceptance of technology, of exercising careful control over what is imported so that essentials of the indigenous culture and social structure remain as the modernization goes on. Licensing technology for existing domestic firms is a great bargain. Know-how agreements and joint ventures are only slightly more costly. But some sacrifice has to be made in any case. The price for modern technology includes coming to terms in some way with those who have that technology in a packaged, usable form. Elites who insist on whipping the dead horses of colonialism and economic imperialism can cause continued stagnation, loss of the foreign-trained technicians, and the realization of the Orwellian vision of *1984*.

The alternative is a flexible policy of borrowing technology at low cost and taking advantage of opportunities afforded. The case of the competent Japanese elites in bringing their country into the modern world and, after defeat in World War II, building Japan again into a modern nation is incisively instructive. It is a model for elites everywhere.

THE SOCIALIST WORLD

With the exception of Czechoslovakia, the countries in the Soviet orbit are or were underdeveloped. The continuing divide between them and Western countries appears to be narrowing due to the United States–Soviet détente. In spite of their historic enthusiasm for socialist self-sufficiency, these countries are slowly becoming part of the world technological circuit. After World War II, a trading bloc (Comecom) was formed, the aim of which was to obtain the advantage of trade within the bloc. Trade with the West was miniscule; typically, bloc countries would buy prototypes of Western innovations and try to reproduce them domestically. But in the 1960's, as growth rates in Soviet countries slackened, there was a gradual movement toward more trade and less self-sufficiency. The contrasting progress of the Western countries and Japan, stimulated by new research-based technologies, was too great, so the self-contained model is being quietly modified. Trade with the West has expanded strikingly. For example, between 1958 and 1968, exports from Germany to the Soviet Union experienced a twenty-fold increase from $12 million to $287 million. Most other Western countries experienced similar increases in trade with the bloc countries, but the United States lagged because of trade restrictions under the Battle Act. However, these restrictions have been relaxed under the last renewal.[5] Although these exports are still a tiny relative proportion of total exports of the Western countries, considerable technology has been and will be transferred to socialist countries under this expanding East-West trade.

Of still more significance in terms of technological transfer has been the increasing development of investment in joint ventures and partnerships between Western companies and state enterprises in socialist countries. Hundreds of such arrangements are reported. Yugoslavia, always the pioneer of liberal socialism, has issued an invitation to Western companies to invest up to 50 percent in Yugoslavian companies.[6] These joint ventures may be for joint development of some important natural resource as in the case of Russo-Japanese developments in Siberia (e.g., timber, oil, natural gas). These projects are typically self-liquidating arrangements in which capital, technology, and consumer goods are supplied by Japan, with returns on investment taken out of current production of the raw materials.[7] Many manufacturing agreements have been reported in Eastern Europe, for example, between Poland and West Germany. Franchising and merchandising agreements involving companies like Hilton and Hertz are planned or already in existence. American companies frequently operate in bloc countries through foreign subsidiaries.[8] These investments and trade agreements, however, are nonpolitical in character, the economic factor usually being stressed, contrary to conventional wisdom.

The reverse policy is pursued where socialist countries are donors of technology; that is, political considerations are usually paramount in providing technical assistance to "the middle world," as the Soviets term them. As the Aswan Dam, an early symbol of Soviet aid, nears completion, everyone is made again aware of the socialist world as an alternative source of technology for development. That the socialist countries have created a place for themselves as a transmission belt for technology to those less advanced countries there can be no doubt. For this they merit inclusion in the circuit diagram of the world system, but it is still true that much of this technology is secondhand, usually older practice bought or learned painfully a few years before. Sutton's careful study documents an enormous transfer of Western (mostly American) technology which aided virtually all (95 percent) the industries of the Soviet Union during the period before 1930, and quotes ruefully from a Soviet source: "In America, they do not guard manufacturing secrets so jealously." [9]

Lenin once made the point that the capitalists would hang themselves with their own rope, namely, their desire to sell to the socialist countries. Perhaps so, but the socialists themselves would be more likely to hang if as certain developing countries are doing, they refrain from obtaining Western technology because of doctrinal considerations. The logic of their position, like that of any technologically backward industry, is to come to terms with the sources of advanced technology and to obtain it as advantageously as possible. This is the course to which they now appear to have returned, with Western business as eager for the hangman as their forebears.

NOTES TO CHAPTER 6

1. U.S. National Bureau of Standards, *Technology and World Trade,* NBS Publication 284 (Washington, D.C.: U.S. Department of Commerce, 1966).

2. D. L. Spencer, "Japan's Pre–Perry Preparation for Economic Growth," *American Journal of Economics and Sociology,* XVII (January, 1958), pp. 195–216.

3. Gunnar Myrdal, *An International Economy, Problems and Prospects* (New York: Harper and Row, 1965). However, the losses from the loss of talent may not be as great as believed. See Harry Johnson in Edwin Manfield (ed.), *Defense, Science and Public Policy* (New York: Norton, 1968).

4. Cf. Dr. William Price's statement concerning analogues with big industries and big organizations, including the military in the United States. Daniel L. Spencer and Alexander Woroniak (eds.), *The Transfer of Technology to Developing Countries* (New York: Frederick A. Praeger, 1967), p. 194.

5. *Battle Act,* renewed with provision for relaxation of control enabling American business to join the brisk business with the East, see Chapter 4, footnote 1; also *Business Week,* December 27, 1969, p. 59.

6. *Time,* April 1, 1967, pp. 11, 29. Throughout the bloc, several hundred joint ventures and partnerships between communist and private enterprises are reported, *New York Times,* January 16, 1970, p. 73.

7. John P. Hardt, *Economic Insights on Current Soviet Policy and Strategy,* Strategic Studies Report 92 (McLean, Virginia: Research Analysis Corporation, December, 1969), p. 43. This study is an excellent appraisal of Soviet attitudes

8. *New York Times, op. cit.*

9. Antony C. Sutton, *Western Technology and Soviet Economic Development 1917 to 1930* (Stanford, California: Hoover Institution on War, Revolution, and Peace, 1968), pp. 318, 319, 348.

7
JAPAN, THE AVID MODERNIZER

The spectacular success of Japan's recovery, remarkably high growth rates, and emergence as a leading industrial nation have become deservedly famous. Before World War II, Japan was ranked as an underdeveloped nation because of its low per capita income and the backward state of its industries. However, informed opinion knew better; W. W. Lockwood's appraisal of Japan on the eve of World War II as being midway between the developed and the underdeveloped nations of the world is more accurate.[1] Today, the appraisal of Japan is swinging in the other direction, as such enthusiasts as Herman Kahn label Japan as the world's most achieving society and predict that the "twenty-first century may well be the century of the Japanese." [2] This is characteristic of the overestimation phase of the cycle with which world opinion oscillates about Japan. Yet Japan's hard-working population is the first non-Western people who, to use Veblen's words, have grasped the logic and hard-driving efficiency of Western industrialism and who also have a goal of continued modernization.[3] For these reasons Japan is instructive for both developing countries and the developed nations.

As in any other nation, developments in Japan are a complex of many factors, but what stands out even on casual examination is its postwar technology policy. In simplest terms, this is a discriminating policy of borrowing technology or technological systems whenever these appear more effective than the old Japanese system. This policy is changing today as Japan's leaders become more aware of the need for indigenous research and development.[4] But until recently, the Japanese policy was simply to borrow the technology intelligently and efficiently. For one illustration, the American military presence in Japan during the postwar period provided a distinct demonstration effect and opportunity to borrow through its management-oriented, research-based technology which had defeated Japan. As Japan had done on previous occasions, a large scale take-over of the foreign system occurred.[5] Beginning as humble and slavish imitators, the Japanese took the latest technology and made it an instrument of home production and exports. Gradually they absorbed and made it their own by improvements and additions until often the Japanese product was the best in the world. Furthermore, though the Japanese demonstrated remarkable flexibility in bringing in the new systems, they were able to preserve the ongoing Japanese way of life in essential ways which were not threatened by the influx of innovations.

FOUNDATIONS OF MODERN JAPAN

The nineteenth-century emergence of Japan as a modern industrial nation has been cited as a classic example of challenge and response. The Western nations, having built empires throughout the rest of the world, turned their attention to "opening" Japan. The Japanese elites bowed with the storm, made certain internal reforms, and proceeded to develop a military force strong enough to defend Japan against encroachment. The Japanese ruling class saw that Western industrialization was essential to the building of a military force. Therefore, the whole package of Western technology and its business system was imported and fitted into the Japanese society. The Japanese had a precedent for such massive borrowing. A thousand years before they had imported and assimilated the cultural system of Tang China.

In the nineteenth century the Japanese were endowed with a very receptive base for borrowing. The closed society of Tokugawa Japan had gone through a transformation parallel to that of Western society from the sixteenth to the eighteenth centuries. Japan went through a commercial revolution similar to that of Western Europe, though unlike Europe political power did not pass into the hands of the merchant class. Toward the end of the era, there was great discontent in Tokugawa Japan, and the advent of Westerners simply permitted previously thwarted trends to be realized. To paraphrase the poetic words of Latourette, Japan was a chrysalis which was strangely ready for opening.[6] Of particular interest is the strong entrepreneurial tradition in Japan manifesting itself in myriad small enterprises side by side with big organizations.

However, these foundations of traditional Japan are dwarfed by the remarkable achievements of the Japanese in acquiring modern industrialization in a period of thirty-five years, ending with the Russo-Japanese War. During the nineteenth century, Japanese development was spearheaded by the need to equip modern fighting forces, especially a navy. It was a narrow, power-oriented, industrial development which left much of the economy rural and small sector, but it still provided an industrial base not far different from that of other mature industrial economies of the day, with the exception of the United States, which was already moving into the stage of Rostow's high, mass-consumption economy.

The nineteenth-century transformation of Japan had certain favorable or fortunate elements. The disciplined, hard-working population had been integrated and made homogeneous over a long period. The elites that existed or were "restored" were highly principled men, used to thinking in terms of national welfare and destiny. Japan entered into world trade when the regnant idea of free, multilateral trade prevailed. Silk, her leading export, was an excellent earner of foreign exchange. Other factors propitious to Japan included the disruption of silk production in France and Italy due to a silkworm disease, and the opening of the Suez Canal.

But however lucky Japan may have been due to heritage and world circumstances, it is difficult to overestimate the remarkable capacity of the Japanese for sheer imitation. Ranking first is their readiness to import foreign technology and second is their ability to reproduce it in exact detail. This propensity to import and utilize technology has often been the subject of caustic comment by other nationals, but it is forgotten that what begins as a cheap and uninspired copy may later be improved so much that it surpasses the borrowed original. In 1869, Okubo Toshimitsu, the first Prime Minister of Japan after the Restoration, is recorded as being in despair over the ineptness and inadequacy of the operations of a textile mill that he visited.[7] In 1962, the Japanese automated-continuous-spinning process, developed by Toyo Spinning and Howa Machinery Companies, was sold to Saco-Lowell, who in turn was licensed to manufacture and sell the process in the United States. The process, which increased productivity 72 percent and reduced crews 42 percent, was hailed by the president of Saco-Lowell as the greatest innovation in textiles since the long-draft spinning process came from France fifty years earlier.[8] This pragmatic method of exact imitation and thorough acquisition of details was central to the success of both the nineteenth- and the twentieth-century miracles of Japan.

Of course other factors accompanied the narrowly defined technology. The Japanese were intuitive practitioners of the systems concept of technology which has been advanced in Chapter 2. Changes in banking, finance, corporate organization, and commercial law were made to conform with Western practice. The Japanese made even more basic changes, such as the abolition of the samurai class to make men equal before the law and in opportunities for gaining personal wealth and advancement. There was literally nothing in the old society which was not scrutinized for sacrifice, if necessary. A minister of education actually recommended discarding the Japanese language for English.

The Japanese, being human beings however flexible, naturally objected to many of the Westernizing changes. The samurai class, bitter over their loss of status and their new poverty, broke into open rebellion. But the despised conscript army, itself an innovation copied from the West, put down the revolt. The beloved General Saigo, today a national hero, committed suicide, as did countless others of the samurai class, but the Japanese closed ranks and continued importing still more technological systems from the West, and eventually they challenged the West with the West's own weapons.

The willingness to bring in innovations must also be set beside the ability of the Japanese to conserve their system as a going concern. Western commentators have long noted the innate conservatism of the Japanese businessman who leaves a Western-type office, wearing his Western-style business suit, returns to his Japanese style home, and changes to his kimono before enjoying a dinner of Japanese food and beverages. In many other ways the traditional system remains viable even under the homogenizing impact of modern technological society.

The visible mechanisms of the nineteenth-century transition are fairly well known. The Japanese government sponsored the industrialization, and when some degree of viability was achieved, the new plants were sold to the private sector.[9] Plants were bought and machinery was installed with the assistance of foreign technicians who were retained as long as necessary, and no false nationalist pride eliminated them before the Japanese could do the job. Yet systematic attempts were made to move Japanese technicians into the jobs as soon as the Japanese were sufficiently trained. For example, in the merchant marine, foreign officers and engineers were gradually replaced with Japanese over a twenty-odd year period, 1882–1903.[10] The Japanese also traveled abroad on missions to study the best practice in each field, and they absorbed what appeared most appropriate and efficient for them. Thus, Britain was the model for the new Japanese navy, the army was patterned after Germany's, their educational system was borrowed from France, and from the United States were borrowed developments in forestry, agriculture, and ways of improving remote areas like Hokkaido. Japanese studied, served long training years in Western countries, and dutifully returned to Japan. As new industries were introduced, this process was repeated. Thus, the United States Strategic Bombing Survey after World War II noted that the top Japanese aeronautical engineers and designers were graduates of such American universities as Massachusetts Institute of Technology and California Institute of Technology, and had served long apprenticeships in Curtis-Wright, Lockheed, Boeing, and other American firms.[11]

Not so well known is the supporting infrastructure which the Japanese developed, based on Western models. Consider the educational system. Although the Japanese retained their own language, they introduced universal education with the result that their level of general literacy today is higher than that in the United States. Good schools were established in every technical field with able foreign teachers. "Boys, be ambitious," admonished the American professor who set up a famous school of forestry in Japan, and his words still guide the forestry service. All of these schools which had incorporated the best traditions of their European counterparts have been greatly expanded after World War II in order to provide for the modern, mass-based type of economy.

POST WORLD WAR II TECHNOLOGICAL BORROWING

The postwar wave of technological induction was based on the existing industrial system, which although narrow, outmoded, and military-oriented, was an excellent foundation for the modern, mass-based system which the United States had attained. However, there were many similarities to the old Japanese miracle of fast absorption of foreign technologies and techniques. Japan's severe wartime defeat, like its nineteenth-century colonial threat, also motivated the Japanese to master the nature of the techno-economic system which had caused the debacle. Again, there were such propitious elements as the demonstration effect of the U.S. military forces and the benign Occupation of the American Forces with their willingness to supply technical know-how virtually free of charge. The Korean War and the aftermath of the Cold War both gave Japan the opportunity to regain political independence and to restore her industry under the impetus of war demand. But the biggest opportunity was technological. The Japanese were handed the new American technology on a silver platter. They were forced to make changes in their old system to meet the needs of the United States armed forces, and thus the foundations of the modern technological system were laid.

THE MILITARY PREPARATORY SCHOOL. The presence of the United States military forces was basically a living demonstration of productive machinery and the famed management-oriented system of getting the job done according to high quality specifications. This operated in various subtle ways, and the ramifications spread virtually through the entire field of economic activities.[12] For instance, the Japanese saw power equipment for earth moving, stored the knowledge, and when their economy had improved, introduced the equipment into their own system.

Of immediate importance were procurements. To supply the American forces in Japan, and later to supply the military requirements of the Korean War, Japanese industry received American procurement contracts requiring them to produce to the same high level specifications, quality control, and a host of higher level practices of American industry. As contractors faced severe rejection rates if compliance was not forthcoming, Japanese industry rose to the occasion and thereby benefited from the chance to learn the American system, to obtain new equipment, and to receive instructions in its use.

The continuance of the United States military assistance program after the Korean War resulted in even larger transfers of technology from such research-based American industries as aviation and electronics. The transfer of advanced military aircraft like the F-86 and F-104 included the plan to give Japanese industry the capability of building and maintaining these aircraft. The effect not only revived the hitherto defunct aircraft industry and transformed it from propeller to jet engine technology, but had far-reaching influence on supporting Japanese industries. The huge military assistance training program in support of such contracts gave thousands of key Japanese engineers and technicians the same training as their counterparts in the United States.

COMMERCIAL CONTRACTS. The military phase initiated and supported a series of commercial, technological contracts, starting from a few in 1950 and gradually snowballing to the present average of more than five hundred a year. Some contracts are in conjunction with investments in joint ventures, but usually they are licensing agreements in which the technology is made available to the Japanese on a royalty basis, often a percentage of sales. In a survey made by this author, the average return to the licensor for a patent license is about 1 or 2 percent and for know-how agreements, about 5 percent. For advanced technology like computers the cost may be higher.

The Japanese began advertising in the early 1950's for the types of technology they wanted to import, listing in a series of bulletins the fields in which they were interested. Foreign companies which responded were referred to appropriate Japanese industrial firms. At first, the Japanese were not very discriminating, except for granting priority to industrial technology and discouraging consumer products of such small consequence as soft drinks or cosmetics. From the start the Japanese government screened the contracts, approving all agreements only after considerable study (and delay). Criteria centered on the productivity of the new technology and its probable contribution to the balance of payments position, either by developing new exports or providing a substitute for imports.

Since 1950, nearly 10,000 major licensing contracts have been approved, touching virtually every industry in Japan and concentrating in the three areas of electrical machinery equipment or electronics, chemicals, and machinery contracts.[13] In the early years the United States was the principal source of the imported technology, but its relative share has declined in recent years, although it still accounts for 60 percent or more of the technology imported annually by Japan.

However, a bare catalogue of events is inadequate to depict the vitality and growth of the Japanese borrowing throughout successive stages of the postwar period. The author interviewed the engineer who introduced the technique of automatic welding to Japanese shipbuilding. Automatic welding was developed in the United States under pressure of World War II needs, and the apparatus was manufactured by the Linde Products Division of Union Carbide Corporation. With the permission and encouragement of the American military authorities governing Japan during the Occupation, this engineer secured clearance to visit the United States and Linde Products Division, where he purchased several automatic welding machines and received several weeks of free instruction in their use. After returning to Japan, he held seminars to teach his fellow engineers what he had learned, and put the machines into operation. With this engineer's help and that of his new contacts at the Linde firm, the Japanese obtained a license to produce welding machines of their own. The phenomenal growth of the Japanese shipbuilding industry to its dominance of producing more than half of the world's ships annually can be ascribed to many factors, including the use of American assembly-line techniques which the Japanese learned from the aircraft and automobile technology transfers. However, Japan's growth as a shipbuilder is inconceivable without these automatic welding machines. Edwin M. Hood, president of the American Shipbuilders Council, stated:[14]

> It is not generally recognized that the techniques employed by our shipyard counterparts in Japan are, for the most part, purely and simply adaptations or extensions of concepts developed in this country.
>
> The Japanese have been able to capitalize on these concepts because of a patronizing government, favorable labor rates and work rules, labor financing incentives, a high degree of standardization and thereby a satisfactory, continuing volume of business.

The transistor is another example of seminal licensing without which the Japanese balance of payments position would be much less healthy. Just a short time after the transistor had been developed by Shockley and his associates at Bell Laboratories, Inc.,[15] the Japanese acquired the basic patent licenses. The transistor radio had been developed in the United States by American manufacturers, but its cost was high. Soon, the Japanese were making the famous little radios and flooding the American and other markets of the world with them. However, as Japanese interests argued before the United States tariff commission, they did not supplant American devices in an existing market as much as they created a new market for "personalized listening," as they described it. American manufacturers, concerned with space and military electronics, simply had not developed the transistor radio for the commercial market. It was not until 1961 that American production of transistor radios overtook and surpassed that of the Japanese.

The Japanese were able to take advantage of rising world trade levels, led by the huge American market. Thus, their technological borrowings were profitable as soon as they went into production. The Japanese with their newly licensed technology turned out wave after wave of new items for export. After 1955, they were exporting secondary textile goods and ships; after 1957, transistor radios and cameras; after 1962, synthetic fibers; after 1963, motorcycles, television sets, and automobiles; after 1964, automotive parts and machine tools; and after 1965, steel, transport equipment, and electrical machinery. Predictions in the 1950's that the Japanese plans for exports of $5 billion could never be attained have been mocked by current exports of $10 billion, and plans call for doubling this figure in the 1970's. This is largely the fruit of a cumulative total of 4,000 licensing contracts estimated to have cost the Japanese about $1 billion over a fifteen-year period.

Despite all this borrowing, it is significant that relatively little foreign capital investment accompanied the licensing of technology. In the early days, American industry with some exceptions, notably the oil companies, showed little interest in investing in the Japanese market. But as Japanese industries began to grow so miraculously, American companies established joint ventures and minority stockholding. However, the Japanese government has been reluctant to approve American managerial control of their industry. At present, the Japanese are under pressure to adopt a more liberal policy. Japan is a signatory to the General Agreement on Tariffs and Trade (GATT) and Organization for Economic Cooperation and Development (OECD) conventions, and the other countries which have signed have accepted proportionately a much larger infusion of foreign (i.e., United States) capital. The Japanese are ardent proponents of free trade and seek investment opportunities in many countries, so that world opinion is forcing Japan to recast its policies. The Japanese like to proclaim that they are bracing for the invasion that they feel will follow the "investment liberalization." However, in spite of claims to the contrary, little liberalization in permitting foreign investment has occurred.[16]

LESSONS FROM JAPAN

There is a school of thought which believes that each country's problems are so unique that little can be learned from another for application in general. This may be true in some cases. But in the transfer of technology, there is sufficient similarity to what is being transferred and in the problems it creates which are independent of the local conditions. Japan's program of simple imitation and the seizing of opportunities is commendable. By pragmatic licensing and by applying the new, licensed technology to the export-oriented industries, the Japanese have regained a prominent place among the world's industrialized countries. Having built up their industries, they are also concerned with increased indigenous research and development efforts, seeking to lessen their dependence on imported technology.

To the developing countries the Japanese willingness to import technological systems in entirety may seem extreme, but it is effective, and the Japanese still manage to preserve their own culture. The Japanese ability to seize opportunities when presented has already been noted by other countries. Korea and Taiwan are seeking to divert procurements for the Vietnam War to their economies in imitation of the Japanese procurements of the Korean War period, and they are succeeding. The eagerness with which Korean workers seek to go to Vietnam for work on construction projects is matched only by the mounting totals of remittances which come back to Korea. It is significant that these two economies, taking leaves from Japan's book, are among the most promising of the countries seeking to break out of the poverty of underdevelopment. Hong Kong too is following Japan's example, and is accused of "piracy" by Japan, as was Japan itself prior to World War II.[17]

It is characteristic of modern Japan to seek and obtain the latest technology from private Western companies on the best possible terms for herself. This, however, does not downgrade the efforts of government-to-government assistance, productivity team visits, students studying abroad, and a host of other channels of technology transfer that Japan also used during her earlier modernization periods. But any country that wants to join the dynamic, growing, modern world economy with competitive exports may well study the Japanese example, especially their willingness to pay the price of entry. Also important was the fact that the Japanese imported only the most up-to-date, most capital-intensive technology at a time when Japan was supposed to be a labor-intensive economy. The Japanese also often had a close affiliation with large American or other Western firms. Frequently, as in the radio and television industry, the Japanese firms became parts suppliers for multinational business firms. The low price the Japanese paid for importing technology has harvested great yields.

Now that Japan has joined the ranks of the rich, developed countries, the latter can profit from the example of Japanese pragmatism. Few shouts of "technological gap" come from Japan. The Japanese have always been too busy closing the gap between them and richer countries. Part of the Japanese success has resulted from stationing their own industrial representatives in every field in the developed countries. Thousands of these "antennas," as the Japanese call them, are on the lookout for the latest technological developments and quickly relay information back to Japan. These "antennas" have been characteristic of the Japanese method of operating. The Japanese rule of thumb is that anything which comes out of the United States research and development pipelines is in Japan six months later.

Though the Japanese have a great phobia about foreign penetration, they have skillfully managed to take over the best technical practice from abroad, and they often perfect and improve it until it is the best in the world. Ships, motorcycles, and cameras come to mind as such examples. The Japanese are prepared to work in cartel relationships with big interests abroad as they have indeed done since the 1890's.[18] Moreover, they are prepared to penetrate other countries' markets with investment. Mitsui Trading Company's establishment in New York to "assist American exports" is already doing $1 billion worth of sales. The Japanese flexibility and open-mindedness toward foreign innovations (but not foreign investments) is in refreshing contrast to the rigidity of governments of other developed nations which will not make the adjustments necessary to keep up with technological progress.

NOTES TO CHAPTER 7

1. W. W. Lockwood, *The Economic Development of Japan* (Princeton, N.J.: Princeton University Press, 1954), pp. 79–80.

2. "Herman Kahn's Thinkable Future," *Business Week,* March 11, 1967, p. 116. Herman Kahn and Anthony J. Wiener. *The Year 2000* (New York: Macmillan, 1967).

3. Thorstein Veblen, "The Opportunity of Japan," *Essays in Our Changing Order* (New York: Viking Press, 1943), pp. 248–66. His idea has been expanded by this author; see D. L. Spencer, "Japan's Pre–Perry Preparation for Economic Growth," *American Journal of Economics and Sociology,* XVII (January, 1958), pp. 195–216.

4. Organization for Economic Cooperation and Development, *Reviews of National Science: Japan* (Paris, 1967), pp. 35–36, 151–55, et alia.

5. For details of the impact of the U.S. military presence in Japan, see D. L. Spencer, "An External Military Presence, Technological Transfer and Structural Change," *Kyklos,* Vol. XVIII, Fasc. 3 (Basel, 1965), pp. 451–74.

6. Kenneth S. Latourette, *A Short History of the Far East,* (Rev. ed.; New York: Macmillan, 1951), p. 260.

7. "I went to see the Iso spinning machine; the way it operates is marvelously smooth and delicate, and no words can describe it. What a difference there is between the intelligence of foreigners and our [so that] we must sigh with shame."—Herbert Norman, *Japan's Emergence as a Modern State* (New York: International Secretariat, Institute of Pacific Relations, 1940), p. 127.

8. *Washington Post,* March 9, 1962; *Daily News Record,* August 17, 1962; and *American Metal Market,* August 30, 1962.

9. Thomas C. Smith, *Political Change and Industrial Development in Japan: Government Enterprise, 1868–1880* (Palo Alto, Calif.: Stanford University Press, 1955).

10. Imperial Japanese Commission to Louisiana Purchase Exposition, *Japan at the Beginning of the Twentieth Century* (Tokyo, 1904), p. 754.

11. Lockwood, *op. cit.,* p. 331.

12. Daniel L. Spencer, "The New Technology in Japan," *World Affairs,* 132 (June, 1969), pp. 13–27.

13. Daniel L. Spencer, "Japan's Industrial Concentration and Technological Pattern in Secular Perspective" in Edgar Salin (ed.), *Notwendigkeit und Gefahr der Wirtschaftlichen Konzentration* (Tübingen and Basel: J.C.B. Mohr, 1969), pp. 242–64, Appendices V and VI.

14. "Japan Capitalized on American Ideas," *Asahi* (English language ed.), February 4, 1967, p. 7.

15. Richard R. Nelson, "The Link Between Science and Invention: The Case of the Transistor," *The Rate and Direction of Inventive Activity: Economic and Social Factors,* A Report of the National Bureau of Economic Research, New York (Princeton, N.J.: Princeton University Press, 1962), pp. 558–83.

16. "Industrial Review of Japan '70," *Nihon Keizai Shimbun (Japan Economic Journal),* December, 1969, pp. 28–29.

17. "Mission to Probe Piracy of Japan Designs in Hong Kong," *Japan Times* (Tokyo), February 1, 1967, p. 10.

18. That the new technology has furthered the concentration of industry in Japan is also part of the price paid. See D. L. Spencer, "Japan's Industrial Concentration . . . ," *op. cit.*

III

CHANNELS AND CONSEQUENCES OF WORLDWIDE TRANSFER

The foregoing chapters have discussed the nature of the technological bases of the modern world. The channels of technological transfer which relate the components to each other may now be examined to gain a dynamic understanding of their operations. These channels provide a conduit through which technology is constantly moving. Some channels are obvious and easily identified; others are subtler and require more specification. The channels are interrelated and to some extent complex. While this brief account can make no pretense of being comprehensive, some classification and description are appropriate to set forth more objectively what is happening, and to provide the first step to consideration of consequences and improvements. However, improving the system is likely to be a complicated program because of the diverse interests and multifaceted nature of the existing network.

8
NARROWING THE GAP: GOVERNMENT-TO-GOVERNMENT

When any social change is contemplated, the first candidate for executor of this change is government activity. Perhaps it is a survival of the feudal age when people looked to the king and patron to take care of their problems, but there is a kind of universal expectation for governments to solve problems. In this case the reaction is: If the technology is lagging, let the established government "do something about it." This is sound instinct and has historical validity, as in the Japanese case. But examination of such cases shows that while governments may initiate action, the accompanying effort of individual enterprise and large institutional change must take place. Moreover, the governments contemplating change often identify the technological induction first with acquisition of skills and second with acquisition of capital. While each of these elements is important, taken individually, each alone is not likely to be adequate. As has been suggested previously, technology must be treated as a system, and its acquisition must be achieved in as close to package conditions as is possible. Segmentation of the problem has led to some of the present difficulties such as strident accusations.

Obviously it is possible to cover only some selected aspects of a large subject. In particular, this chapter will seek to describe the role of government under categories of person-to-person transfers, the stimulus of the United States government's program, what other governments can do, and conclude with a proposal for an International Technology Agency to deal with the complexities of technological transfer. The first area is selected for highlighting because the evidence is accumulating that transfer of technology is done effectively through face-to-face confrontation. People learn from reports in printed media, but individual relationships are essential. The second and third areas are logical first approaches. The fourth section considering an international agency follows from the multisided nature of the problem.

PERSON-TO-PERSON TRANSFERS

For an underdeveloped country taking its first steps toward modernization, the logical thing is to send abroad a team of high-level men to observe what is happening in certain essential fields, and how feasible arrangements may be made for transferring technology in a given field to the home country. Stemming from the initial mission, selected students will be sent to study abroad, and foreign technicians flow in to lay the technological groundwork for students when they have completed their studies. Following the Japanese pattern, elites who control the government and are sensitive to the need for change initiate the process by selecting large-minded men to make the initial mission and able, nationally dedicated young men and women to follow up. Thus, a mission from the marketing board of a small African state came to Washington, D.C., some years ago with a requirement for training secretaries and marketing personnel. A reliable and energetic American lawyer was engaged to act as agent, a university was approached to carry out the education program (it subcontracted out those parts it couldn't provide), and contacts with the Agency for International Development (AID) to finance the package were initiated. All sorts of difficulties have been encountered, but the plan is gradually being realized. If the temptations of lengthy United States living do not overcome the dedication of the students to their country's needs, the marketing board's program for liaison with the modern world is likely to be effective.

The inspiration is usually the greater in the first generation of students. If they are impressed with the mandate from their elders, they are likely to return; and given proper institutional implementation, they multiply the results of the training many times in schools and other training programs. Some of this training will be diluted by the natural tendency to hold back the expertise in order to preserve one's original position as the authority; but the important things will be passed on, especially if reinforced by the foreign technicians whose role, initially monitorial, may, for further innovations, become progressively consultative, as in military follow-on agreements. A big danger is training students haphazardly without clearly defined goals and complementary investments of other ingredients in the technology package. It is in this unprogrammed study that students lose their sense of mission, and are likely to be lured away in the brain drain.

Precise programming for foreign students is difficult; the government is not the only sponsor of students. Many come under private sponsorship of family, foundation, or by their own efforts. Many return and give excellent yield on the investment, but many times they become "brain gains" of the developed country. The educational training they pursue may indeed have little relevance to the home country's needs, and in terms of alternatives, the individual may really have nothing else to do.

The underdeveloped country must establish its own educational system as an effective training instrument and gear it to the specific needs of technological transformation. Nationals who are trained at home who move into the new local technologies avoid the problem of brain drain. Some restriction on the open flow may be considered in the future. International agreements may help developing countries to work out these policies, for, on its side, it is the responsibility of the developed country to avoid subsidizing its own manpower planning program inadequacies with the ad hoc coverage of the foreign trainees.

DEVELOPED COUNTRIES. For the developed countries and those underdeveloped countries beyond the first development phase, the government's role in person-to-person transfers is more complex. Not only must it try to refrain from subsidizing its skilled labor supply at the expense of the underdeveloped countries, but it must set up mechanisms to prevent the slipping away of its own talent. Education of a scientist or engineer is expensive, and the investment must be safeguarded. The developed country's government should be certain that the educational system is itself an effective instrument for coping with the modern technological world. From time to time, high-level commissions must make surveys of the country's needs, examine educational policy in other advanced countries, and make recommendations for improvement. Sharp updating and reorganization of education have followed from acting on such recommendations in the United States as the Flexner Report on medicine[1] before World War I, and the Gordon and Howell survey of business education after World War II.[2] One of the most striking in recent times is the Robbins' Report on Higher Education in Great Britain which has resulted in broadening the educational base and which is making the higher education system more responsive to the needs of the new technology.[3] As the new technology inundates us, such studies must come with greater frequency.

After insuring an adequate and incisive educational base, the government's role is to plan for adequate utilization of the graduates. Manpower planning programs for quantity and quality of jobs must be utilized. Unemployment is an obvious cause of emigration, but this has not been a problem in most developed countries which apply the fiscal and monetary policies of modern economics.[4] Aggregate full-employment policies may not insure against recessions in particular industries as the result of government budgetary decisions. The loss of aeronautical engineers from Great Britain resulted from the cancellation of government contracts.

In spite of such work stoppages, the big loss of talent has often resulted from dissatisfaction with either the conditions of work or the salary. Conditions of work include many nebulous as well as specific matters. Participation in the management of the laboratory may be quite as important for an independent, scientific type of person. American "think factories" try to make conditions similar to a university department or faculty with much shared decision making. The corporate structure includes a large proportion of managers and directors with scientific backgrounds. Technical men feel better understood and wanted in such an environment. Then, the physical artifacts of equipment are very important to research. The scientist may feel frustrated when purchase of a valuable piece of equipment is blocked by the corporate hierarchy for budgetary reasons, or by the government hierarchy for balance of payments reasons.

But the biggest cause of emigration is undoubtedly the salary and combined fringe differentials. Scientists and engineers, no matter how dedicated, are still faced with the problems of supporting themselves, and like other men in other areas, they will tend to gravitate to higher income levels. Even when allowance is made for the higher cost of living in the United States, the income is often two or three times the level in England, and even greater compared to other countries.

If it is deemed desirable, what can be done by the home government to stop this kind of drain? If the educational system is turning out more scientists and engineers than it can use, their utilization in the United States is a blessing. In an equilibrium of world flows of trained persons and technology, what has gone out as trained brains will return as licensed technology after some lag. Nothing is lost from a global point of view, and if time lags are disregarded, there will be gain in the 1970's. But there is the problem of technological nationalism. The national entity is a going concern, and it is important not to fall too far behind or to be dependent on other nations. Over a long range and with the help of international agreements and more international thinking as suggested below, the national entity may not find this as thorny a problem as it now seems.

In the foreseeable future, however, the national government can take some ameliorative measures: the establishment of a ministry of technology which will grapple with the problem of the emigrating engineer and seek to improve his employment alternative at home is an obvious course of action. If the government feels that technology has a priority in its national goals, many things can be done to improve the potential emigrant's status. Business can be cajoled by "carrot and stick" to involve him in management. The laboratory equipment purchase which has been blocked can be obtained with government expediting. The salary differential may not be bridged entirely, but much can be done in tax rebates, expense accounts, housing, and other concealed subsidies to a man's income. A government itself can set the standard by paying higher salaries to its personnel. A full matching of the overseas offer need not be made, for there are tangible, psychic benefits for most men in not moving. Similarly, reverse drain operations may be undertaken with such incentives as paying the man's moving expenses to return home, an item that could be an allowable deduction on corporate tax returns.

But these palliative measures cannot take the place of the government's sponsorship of large programs of research and development, and other more basic approaches. Generally, a large national commitment of resources to an indigenous research effort is essential to provide technological and scientific opportunities. Making research and development contracts with firms and universities, following the American pattern, is one way to set conditions for keeping scientists and engineers in their home country. The scale of effort is bound to be smaller than that of the United States, but some alternatives will be generated. There are no doubt other reasons, but it would seem significant that the one European country which has mounted a serious technological research effort has the least brain drain. Only in France is there a genuine national effort, and the government seems determined to make it work. Japan's efforts, notably in space, though very small scale, are evidenced by mounting research and development expenditures. It too has little trouble with brain drain problems except in special fields, notably mathematics. Yet it is widely recognized that the size of national income limits a country's ability to commit resources to programs involving heavy research and development efforts. A national space program, for example, cannot be undertaken without the danger of duplicating existing efforts with uneconomic, small-scale experimentation of activities which are being better performed in the United States and the Soviet Union.

A European-wide effort springing out of the Common Market is an intellectually more satisfying vision. Unfortunately, its precedents and prospects are not encouraging. Successful cases of intercountry collaboration in science and technology are few and quite specific, such as the Franco-German A-300B jumbo jet plane. Where they exist, international authorities tend to be underfinanced. The difficulties of integrating the separate scientific and technological traditions and imbuing Europeans with the willingness to commit resources to international bureaucrats who may lose touch with the people paying taxes are great. While these ideas are under study in all the cooperative European organizations, one can hardly be sanguine about launching a determined effort of European technological cooperation. Rather, there is an expectancy that the United States government should take the lead in actions designed to improve the situation.

THE UNITED STATES GOVERNMENT STIMULUS

Basically, this longing for United States-sponsored solutions is undesirable. From one point of view, European governments must get on with the job of creating an integrated economic entity with adequate income levels, which can be taxed to provide the resources necessary for the large-scale programs of modern technological research. Europe has more chance of developing its own potentials, if that is its goal, without United States participation. But if the supranationalism of a European research heartland is foregone, and the vision of the United States as a potential "technological empire" is allayed, there may be no need for such an independent research development. The present system has worked reasonably well to date and may continue to work adequately if some of the chauvinism is alleviated by a few deft adjustments. If not and if the passion for nationalist technology is so great, the best role for the United States is less action, not more.

Briefly reviewing the history since World War II, the United States has poured forth its technology unstintingly, and enormous gaps in every technical field have been closed between the United States and other developed countries. Much of this technology was incorporated in the products which were made available under the Marshall Plan or by even earlier funding. Incorporated in, say, a tractor was the best technology of the latest vintage which influenced the design engineers and technicians of the recipient country's firms. Overt or covert pirating sometimes occurred, but usually a patent or license agreement was secured, and the local firm obtained servicing, overhaul, and manufacturing rights. As European prosperity returned, the transferred technology embodied in the machines was produced in volume, and the improved technology and products were soon part of the everyday life of the recipient nations.

Similarly, many production processes were transferred along with the capital in this postwar period. Many American management teams were sent under government sponsorship to improve the conditions of industrial and agricultural production abroad. Conversely, foreign productivity teams were brought to the United States, and had an opportunity to see the latest practice in American plants, laboratories, and other establishments. Much of the process was less conscious; a good deal of technology was absorbed by a kind of osmosis which happens when a piece of unfamiliar equipment is bought and must be operated and serviced. Government initiative in such ways would often pave the way for establishment of commercial relationships which would grow into longtime business alliances. Often the United States government men making the arrangements were former businessmen who had retired or were on leave from the industry for which they were responsible in the government transfers. United States government policy facilitated in every possible way the acquisition of American industrial and agricultural know-how.

Nor was this done in an empire-building spirit of American business preferment. Generally, the United States government's activities were set on a broad-gauged basis, and while some favoritism of American business interests might have been exercised, other world business interests were encouraged to take part. Thus, in the off-shore procurement programs, many other national industrial structures were given contracts to supply the local areas, permitting them valuable local contacts for later business relationships. For example, Toyota Motor Car Company in Japan supplied many military trucks, jeeps, and other vehicles and parts to other countries in South Asia, giving Toyota invaluable access to markets in those countries. In Taiwan, Japanese business has been built upon many of these earlier connections. In recent years, because of the problem of balance of payments deficits, it has been necessary to favor United States sources of supply to a greater extent, but there are still examples of foreign firms obtaining contracts.

This openhanded policy continues and applies across a broad spectrum of international activities. Thus, in the investment field, American business was formerly quite timid and inexperienced about going abroad, but in recent years the recognition of such growth areas as the European Common Market has resulted in a large outflow of capital. United States government policy has encouraged foreign investment and accompanying technological skills, but the balance of payments problem has been invoked to discourage the flow on a discriminating basis. There are also restrictions based on the United States' antitrust policy, but actually in most cases these work to the benefit of the recipient countries' small businesses. In Japan's case, the original patents on the transistor were made possible by the American government's insistence on unrestricted licensing. There are restrictions on trade with Soviet countries, but even these are being relaxed, as noted in Chapter 6.

Similarly, in line with beliefs in unhampered personal migration in response to incentives, the United States policy has been to accept students and practitioners without much thought to the impact on other countries. There is, it is true, one type of visa which admits people for a specific period of time, and when they leave, they cannot return until they have lived for two years outside of the United States. But in general, little restriction is placed on the choices of people to pursue occupations and come and go freely in the United States. That this policy should create a problem for other countries was not previously recognized.

AMELIORATING FRICTION. Assuming that narrowing the technological gap is a desirable goal for all nations, what in fact could the United States government do to ameliorate conditions? The basic position to be recognized by men of good will who face the future without fear is that the existence of the gap is a constant challenge for all nations; and both developed and underdeveloped economies are feeling the prod of international technological acceleration which, sooner or later, evokes in them the response to modernization. Thus, in the case of the developed nations, the lag in ability to compete with the highly research-and-development-oriented American industries forces them to get on with the work of larger economic unification, such as the European Common Market, which has been so productive; or it forces governments to reconsider their technological policies. It becomes evident that a government must make greater provision of funds for research and development where market stimulus is lacking, or they will fall further behind. In this sense an argument can be made that the United States government should do very little to interfere with this drastic remedy. But the United States government is sensitive to the needs of the rest of the family of nations, though it certainly does not want to play the unmitigated role of international technical catalyst any more than it wishes to be an international policeman.

First, some of the more obvious irritants can be alleviated. Thus, it would probably not be difficult to obtain the cooperation of leading American defense contractors to cease some of their direct advertising for talent from abroad. People in the developed countries are very well aware of the opportunities which exist, and the frontal advertising campaign such as that in Figure I, has been painful to other people in the country who do not like to see the exodus. Working through more such discreet local channels as department heads of university faculties or trade association executives can well accomplish the same or better results for modest fees. Second, if the companies can be encouraged to phrase their offers on a temporary loan or trial basis, the impression will be given that the break with the old country is not permanent and that returning home will be possible. Perhaps a tax concession or even subsidies for personal travel to return home for vacations or sabbaticals would be appropriate. Third, and even stronger, American companies could be encouraged to subcontract more research and development to subsidiaries abroad. Building up these foreign affiliates as centers of research may well have cost-benefit advantages against the present typical company policy of centralizing research and development in the United States. Fourth, the government procurement authorities could establish "small country set-asides" on a long-range basis to provide an encouragement to smaller industry abroad which is similar to that extended to small business at home. Fifth, greater encouragement of cooperative research and development programs

Figure 1. Friction-creating advertising

ARRIVING FOR INTERVIEWS
APRIL 10—19

U.S.A.

opportunities in the telephone industries for Scientists and Engineers

Expansion in U.S. markets has created career opportunities in research, design, development, and manufacturing in one of Mid-America's largest industrial complexes. This company is a leading producer of electronic and electro-mechanical telecommunication switching equipment. This includes automatic exchanges, dial telephones, automatic direction of calls through multi-exchange networks, local and toll switching, and subscriber toll dialing.

Logic Systems Design & Evaluation · Systems Reliability · Electronic Packaging · Circuit Design · Traffic Analysis · Electromechanical/ Magnetic Component Design · Telephone Switching Systems Planning · Transmission.

Manufacturing Engineering: Solve production problems for electronic and electromechanical products, equipment selection, process new products into manufacturing, cost analysis, plant layout, and provide technical assistance to manufacturing departments. Positions exist in the following areas:

Chemical-Metallurgical · Processing · Methods · Test Equipment Design · Tool and Machine Design · Automation.

All positions require a B.Sc. (Eng.) and 2 years experience.

WRITE OR TELEPHONE TODAY

To apply simply telephone 01-584-3768 Monday to Friday, or send a copy of your curriculum vitae (or its equivalent) and your telephone number to: **Mr. Allan Eaton, The Manpower Register, 31 Queen Anne's Gate, London SW1.** Your application will be treated with complete confidence. There are no fees of any kind.

Careers Incorporated (U.K.) Ltd.
The Manpower Register

Note: Nine additional leading USA employers will be here for interviews April 10. If your qualifications do not match the requirements of the employer above, contact the Manpower Register at the above address for consideration by the other nine employers

Source: *Times* (London), March 26, 1967, p. 58.

could be offered, giving such centers of research and development abroad as much opportunity and encouragement as possible. Perhaps this might include their taking responsibility for assuming leadership in well-defined areas of research or development. Such ameliorative steps may do much to promote better feelings among nations and at the same time smooth the flow of technology.

More fundamentally, the United States can make sure that its own educational system is supplying the right quantity and quality of trained engineers and scientists that are needed for carrying out the demands of its research and development. While the explosive growth of research and development expenditure cannot continue at the geometric rate which it has experienced in the last few years, the demand for trained people is still very high and is likely to continue so. This demand is intensified by the stockpiling of scientists and engineers which goes on to a considerable degree in an industry always bidding on contracts and planning for the future. Such practices intensify imbalances and lead to practices of raiding other countries' scientists and engineers.

Basic corrective measures require changes in the educational system to make available a larger number of trained personnel. Therefore, attention must be paid to the motivations of children, beginning at early ages. Inducements and incentives must be set up at every stage to guide young people into areas where they can meet their country's needs. Thus, if the American Institute of Physics' estimate of 20,000 physicists required by 1970 is correct, some inducements should be offered to undergraduates to study physics rather than biology. Though it is hard to predict in advance, long-run measures of this kind will help to prevent raids on the products of other educational systems. Although this will not solve completely the problem of keeping the gap manageable, it would be a desirable action. For as was stressed to the author in many interviews with foreign specialists, the United States government can do nothing fundamental, and it runs risks of wasting its time and money. The problem, in the last analysis, is what the lagging country must do to set the conditions to help itself.

WHAT OTHER GOVERNMENTS CAN DO

Governments of countries, both developed and underdeveloped, have options open to them in responding to the challenge of the modern technology. At one end of the spectrum of possibilities, they may choose policies of isolation, hothouse industry, controlled and regimented economies which segregate their country from the main stream of technological advance. Alternatively, at the other extreme, they may seek out and eagerly induct modern advanced technology and management of the research and development world. Most countries' governments tend toward the latter end of the spectrum, and for them the problem is to set conditions conducive to the introduction of technology. The cultural resistances within a country must be overcome with "carrot and stick" programs devised by clever ministries of technology. This will not be a sufficient answer, but it would seem a prior condition precedent to improving the channels of technological flow and narrowing the gaps around the world.

The task of the governments of underdeveloped countries is more difficult; its prescription, relatively simple. If a developing country wants technology and the elites governing the country have some kind of concensus and control, the sensible model is the Japanese. The technology must be brought in and the indigenous culture modified or in some way displaced to make acceptance possible. Government leadership in the initial stages is essential, and its continuing surveillance and screening for appropriate technologies is necessary. Its leadership role is to obtain technology in the package form on the best possible terms, together with the necessary capital, skill, and managerial talent. It should take full advantage of such opportunities as military activity, a point to be developed in the next chapter. In parallel, the government's role is to establish an educational system which has some relation to the needs of the incoming technological systems. To this end, judicious controls over who gets a foreign education and in what subjects are essential. Similarly, decisions as to what technologies are to come in must be made, but the possessors of most technologies are the large international companies in the Western democracies, and unless the elites are prepared to join the Soviet bloc, they must come to terms with the techno-business package offered by the West. The amazing vitality and growth of poor countries which have followed this technological policy—Korea, Taiwan, Hong Kong, Spain, Israel, and Mexico, for example—evidence the wisdom of this course of action.[5]

The developed country's government technology policy is more complicated. These are rich and comfortable countries, perhaps not quite as rich as the United States but well off, and the need for major change is neither clear nor pressing. A government responsible to the electorate has difficulty in making decisions which seriously affect vocative, entrenched groups. Yet the pace of technological advance is so rapid, and the planning horizon or pipeline of research and education of necessity is so long that decisions must be made far in advance. The problem of continuous response to the technological challenge can be viewed on two levels. There are some fairly simple stopgaps which are open to a government in the short run; in the long run, long-range institutional policies must be introduced which may have important consequences in improving a country's technological position, but at some cost to the established system and, for that matter, to the establishment.

In the short run, one may employ the tax system to create special conditions favorable to the retention of scientists and engineers in their own countries. In a country like England where a third of personal income at professional levels may go for taxes, there is a large cushion to make concessions in favor of the technically qualified which will go far in narrowing the gap between the overseas offer and the present situation. In countries where government scholarships are made available, a condition for acceptance may be the kind of position which must be filled on completion and for how many years. More positively, matching efforts may be made. Sweden, on learning that a distinguished mathematician was about to depart for Harvard University, created a chair for him at a Swedish university with matching salary and emoluments.

On the industrial side, the government can encourage greater efforts to build research and development activity in the country. The most obvious is to let research and development contracts in support of expanded programs in nationally important fields. But there is also a wide range of clever and highly flexible incentive techniques open to a government which wishes to influence private industrial activity. These incentive contracts modeled on French-style indicative planning* can be invoked to channel company efforts into greater self-financed research and development. Borrowed from schemes used to secure other objectives, for example, export promotions, a wide array of incentives can be devised to bestow on firms which will cooperate by investing in research and development efforts to a greater extent than they are doing. Thus, schemes of accelerated depreciation allowances can be used to encourage research investments. Benefits including preferential access to credit, low-interest loans, or even cash grants may be made available with

* French-style indicative planning refers to directing a market economy, in contrast to Soviet planning of a command economy.

appropriate safeguards and deadlines calculated to spur action. Thus, limited periods for application, fixed opening and closing dates, conditional renewals, penalties for failure to produce results, and other schemes may be used. Such incentive and penalty techniques can be worked out by clever government planners in master research and development blueprints for the economy and individual industries.

Another intervention that a government can make is to throw its weight in support of the engineer-technician as against the prestigious, theoretical scientist. Building up the applied side of the sequence from idea to commercial product requires giving more prestige, social status, and income which governments have ways of encouraging. Emphasis on management and marketing is particularly necessary, as these tend to be neglected fields outside the United States. Like the Greeks of old, too much attention has been paid to the basic side and too little to the role of technology in bringing out the final "bug-free" product.

There are government research contracts in support of development programs in most developed countries, but they can be allocated still larger proportions of the national product. Comparisons show that relative research efforts, with the possible exception of the United Kingdom, lag behind the United States. If governments want to overcome lags, still more funds must be allocated. It is also necessary to direct the research into fields where comparative advantage is likely to exist. Further research funding, if it is to be effective, cannot neglect the applied side. One can determine optimums here by looking at the yields from the input, and the general impression is that Western European research does not seem to yield results comparable to American research efforts.

An adequate national technology policy is needed if results are to be obtained. Europeans typically talk of a "science policy" rather than a technology policy, and the term *science* reflects an unconscious bias toward the theoretical side. The term *technology* puts the emphasis where it is most needed. In any case, technology policy includes the management of resources, the creating of conditions for growth, and the obtaining of effective yields for the economy, defense, and other spheres. A modern government's big role today is learning to organize research technology on a national basis, and this is a difficult process.

Adding to the difficulty is the fact that the national scale may be too small to compete effectively with the United States, its huge market, and its large, central government funding of research.

While there are undoubtedly many industries and areas where scale is not important, it is crucial in the cluster of aerospace, aircraft electronics, computer, nuclear energy, and related glamour industries, from which so many creative innovations have been derived. For competition in these fields, some sort of a supranational effort is necessary.

If a government wants to compete in such areas, and most do, it must think in terms of joining some larger geographical area such as the European Common Market. But Europe is not a national entity, and the international collaboration thus far produced has not been spectacular. The examples of such existing programs as the British-German-Dutch uranium enrichment project are in very specific fields, and the magnitude of research funding of such inter-European agencies as Euratom* does not compare with the United States. This condition may not be true in the future, but the differing institutional structures and outlooks have not made for rapid development here. Possibly the proposed British entry, if accomplished, may change this; in fact, the British fundamental proposal bases much of its claim to entry on the idea of a larger technological Europe.

Alternatively, it has been proposed that Britain, as a key country, join an "Atlantic" community of nations consisting of the United States, Canada, Australia, New Zealand, and Japan.[6] (Presumably, the last three countries are thought to have socioeconomic levels and structure sufficiently similar to the Atlantic countries as to make them good partners in spite of their distant locations.) This larger grouping theoretically would attract other countries at a later date, perhaps the European Common Market itself. Still other regional common markets have been discussed as bases for larger technological efforts in other parts of the world. For most of these, the market stimulus is not adequate, and the central executive agency, even in the European Common Market, is not likely to provide the government financial support that it does in the United States.

The uncertain state of this Common Market thinking regarding a larger European technological effort is reflected in current reports. For example, a European market commission examining market technological potentials does not suggest central financing, but rather a series of coordinated national steps. These steps include the kinds of incentives like those noted above for a government acting

*European Atomic Authority.

alone—common depreciation stimulants, patent schemes, company organization, and intra-European movements of capital.[7] Bridging the gap between the idea of larger scale research technology and its finite accomplishment appears to be a difficult, long-term task.

AN INTERNATIONAL TECHNOLOGY AGENCY —A PROPOSAL

These halting proposals for work at the European Common Market level and similar activities of the Organization for European Economic Cooperation and other bodies at the regional levels are commendable, but they suffer from an ad hoc character, a kind of bureaucratic timidity, and the usual regional limitation of exclusiveness. What is needed is a larger, worldwide organization of stature to examine the problems of international technology in a worldwide context, and offer equitable, reasonable courses of action for all countries. Regional analyses may be carried on within such a framework, but the complicated issues and crosscurrents of international technology require a more worldwide vista.

The logic of international technology requires a separate international agency to deal with its complexities. New thinking and new approaches are necessary. The establishment of an International Technology Agency to take its place alongside of the International Monetary Fund, the World Bank, the Food and Agricultural Organization, the International Labor Organization, and other specialized United Nations agencies would be a desirable course of action at this juncture in world technological affairs. A conceptual framework in which to think in terms of the world and its interests as a whole can be achieved only by this sort of an international ministry of technology for the whole of humanity.

The analogy with a technology ministry of a national state is somewhat misleading, because the agency would have operating functions which in addition to ministry functions would involve it directly in the activities of research operations. It is of course impossible to do more than give a bare sketch of a proposal which would have to be formulated in great detail as were its famous predecessors, but some likely cardinal features may be suggested.

Such an agency would have the functions of an international forum and clearing house, a data bank of technological talent and their whereabouts, as well as a catalyst for large aggregates of investment necessary to carry out large-scale research. Like the World

Bank, it would make loans to finance investments and participate actively or passively in the establishment of research centers and projects in member countries. It would derive income from its share of the forthcoming innovations and patented processes, which income would be added to funds available for further research.

Its structure and organization would be patterned on the Bretton Woods Agencies with voting authority in proportion to the contributed capital. It would be given power to raise its own capital in the financial markets of the world alone, or in collaboration with national states, banks, or large companies. In its operating role, it would contract research to firms and companies very much as a national state does. It could undertake this on its own or on behalf of member states. Resultant innovations could be exploited through channels of private business or through government corporations in the member states. In either case, royalties and other fees would come back to the agency as the sponsor. Participating governments would be under no obligation to give preferential treatment to the innovations or to the corporations marketing them, but a system of incentives might be worked out for national assistance to aid the spread of the innovations.

As with the Bretton Woods and other international organizations, much careful thought would precede its establishment, and its functions would be extended progressively. Initially, the clearing house and consultative functions could be stressed. Small countries would find its advice helpful. As the agency gains in stature, it could progressively take on more complex functions. Coordination of national research efforts of several countries or groups of countries might be a next step. Assistance in forming regional research programs associated with regional market blocs would be another useful role.

A trial operating function of sponsoring research could be done on a pilot project basis initially, and when experience is gained, larger scale activity could be undertaken. The forum function would come naturally both in the inception stages of discussion and in the convening of a chartering group as well as in the progress of the institution over time. Forum activity which now occurs in the political press debates or in specially convened conferences would find a sounding board and more effective ways of reaching consensus and action in the International Technology Agency.

The clearing house function would include several components. First, it can be tied to the knowledge explosion and the problem of retrieval of information. A data bank of what technologies exist, for what purposes, under what terms and conditions they are available, and what sort of social repercussions their introduction is likely to cause would be very valuable to potential users. Second, a corps of international civil servants, devoted to keeping abreast of what is happening in the many fields of technology, would provide reliable information for an intelligent world opinion. The kind of independent evaluation associated with the World Bank missions could be developed in this technology field. Setting of standards, as in statistical reporting, of what constitutes research activity in comparative studies would be another such clearing house function.

The provision of expertise for feasibility studies of research areas and topics in member countries is an obvious function, having a precedent in the work of the existing international agencies. Teams of technical men could be made available to countries to provide advice, to write proposals, to organize projects, or to expedite better operations in a country. International commissions of such men under the agency's auspices could study such problems as the Atlantic brain drain and offer objective analysis, equitable compromises, and feasible solutions for all concerned.

The catalyst or operating function would follow naturally from the others. If, for example, the brain drain study group decided to pool scarce international scientists and engineers to be borrowed for appropriate missions, the International Technology Agency would be the logical institution in which to establish the pool.

But it must be acknowledged that investment in research activities would be a delicate and difficult matter. Here, the proposed agency would stand in the place of a national government and would sponsor large-scale research in similar ways. Admittedly, this may be too large an order for the national states to accept, but if the developed countries particularly are interested in cooperative efforts without sacrificing their sovereignty, this kind of organization would be at least as effective as the European cooperative attempts of the present, and would have no more drawbacks. It would be able to initiate research in areas other than defense or space, and thus be potentially a diversification force.

In theory, depending on the size of its mandate, the agency could mobilize resources on a large enough scale to perform almost any research activity, but pilot projects aimed at significant problems in a wide spectrum of knowledge would be a good beginning. Generally, the research might aim at advanced types, where scale and resources were beyond the reach of even the developed countries.

But some significant research problems particularly in the social fields, which are vital to the well-being of the developing countries, might be undertaken. A well-financed, concerted research attack on the population explosion in a country like India, for example, might develop a tested social technology for curing the problem in a generation.

An International Technology Agency would, as its name implies, gear itself largely to a problem-solving approach in the research field. The method of systems analysis which has been employed so effectively in physical and military sciences in solving such problems as spaceships, missile systems, and supersonic transports can be given global proportions by such an agency. But its mission would be closely linked to the member states which brought it into being. It would in some degree compensate for the lack of large markets and big public financing of research in smaller countries, and, therefore, it would have to be responsive to their needs.

Further, the idea would be to create an agency concerned with international technology from a multiregulatory viewpoint. With the new force of technology creating repercussions in many directions, it is only prudent to try to create a stabilizing entity capable of righting balances through world-perspective surveillance. In particular, international self-regulation of business by treaties among countries could be administered by the agency to alleviate the continuing conflicts between national states and affiliates of multinational companies.[8]

Lastly, it would provide a possible vehicle for cooperation with the Soviet world in the technological field. Ideas to explore a program to establish an exchange of management and technical know-how between Communist and non-Communist countries have been discussed for some time. The notion that advanced industrial countries in both worlds have certain problems in common, such as urbanization, transportation, water and air pollution, etc., has received some attention, and the Ford Foundation is reported to have been studying the matter. Creation of such an International Technology Agency as proposed here would provide an obvious forum for opening a dialogue on this and other cooperative East-West matters pertaining to technology.

NOTES TO CHAPTER 8

1. Abraham Flexner, *Medical Education in the United States and Canada,* a report to the Carnegie Foundation for the Advancement of Teaching (New York City, 1910).

2. Robert Aaron Gordon and James Edwin Howell, *Higher Education for Business* (New York City: Columbia University Press, 1959).

3. Great Britain, Committee on Higher Education, *Higher Education, Report of the Committee Appointed by the Prime Minister under the Chairmanship of Lord Robbins, 1961–63* (London: Her Majesty's Stationery Office, 1963) Cmd. 2154.

4. For how fundamental economic growth is in coping with such problems, see Walter Adams (ed.), *The Brain Drain* (New York: Macmillan, 1968).

5. Similarly, the Western agro-business package of transferred agricultural technology has produced the fabulous "green revolution" in food production in India, Pakistan, the Philippines, and other Asian countries. Clifton R. Wharton, Jr., "The Green Revolution: Cornucopia or Pandora's Box," *Foreign Affairs,* 47 (April, 1969), pp. 464–65.

6. Professor Harry Johnson's comments as reported in *The Times* (London) March 23, 1967, p. 1.

7. *The Economist* (London), March 25, 1967, p. 1175.

8. As this book goes to press, this idea of international supervision of multinational companies has also been publicly advocated by Professor Raymond Vernon and a French businessman, Robert Lattes, *New York Times,* January 14, 1970, p. 61. My proposal would be for self-regulatory action similar to labor-management conciliation practices.

9
INTERNATIONAL MILITARY TRANSFERS

In developing a more effective and equitable world technological system, the other side of the spectrum from the International Technology Agency vision is to make use of institutions already in existence and seek to improve them. One such institution, which is a branch of the government channel, though usually given separate identity, is the military. The word *military* does not have a popular, positive image for many persons because its security mission implies violence, but if its role is considered objectively, it will be seen that like all institutions, its functions have evolved over the years and especially during the present. Much new technology is closely related to military activities; fighting is a function of technology. Historically, the military institution has always been a source of demand for innovations, and even more so today. The areas where research is generating new technology today are in large proportion military or military-connected. And perhaps even more significant is that military forces transfer the technology from country to country wherever they go. Closer scrutiny of this ongoing institution of technological transfer with a view to better utilization is rational for even the most confirmed pacifist.

THE MILITARY FOUNTAINHEAD

Stretching back to antiquity, the military has always been a fountainhead of technological innovations which have penetrated the civilian economy. The improved spear, born under the pressure of wartime demand, had effects on the hunting industry. The durable roads developed by old empires for military communications and logistics provided infrastructure for economic specialization, trade, and growth. The industrial discipline of the factory system in the industrial revolution was patterned on the discipline of the military. Military procurement in providing a guaranteed market led to scale production techniques. Development contracts are much older than generally recognized. Interchangeable parts for rifle development were attempted by Eli Whitney for the United States Army around the year 1800.[1] As military operations have become more complex, military training has provided counterpart civilian occupations. In many countries the ex-serviceman's training and skills have subsidized industrial training for civilian use.

The conscript army has been a source of homogenizing, upgrading, and orienting the populations of national states with a larger vision than that of the village. The general awareness of the existence of modern products and ways of doing things is conveyed to the peasant lad, who returns to his civilian existence with new concepts and visions of improvement. The hierarchy, far from stifling the individual, as is popularly imagined, provides a goal of advancement as a reward for initiative. Entrepreneurship and new ideas are often generated in military circumstances. At minimum, some feeling for how large organizations work and the importance of taking responsibility for some aspect of the whole operation is conveyed to large numbers of young men in the population. Civic action programs in less developed areas too have now become a direct responsibility. The vision, so popular in the United States, of a drone-type military confined to barracks life in peacetime is quite the opposite of modern military forces.

The direct economic role of the military forces in engineering bridges and roads and damming rivers has long been taken for granted. So too has the military role in flood relief, coping with disasters, and other large-scale emergencies. Napoleon's soldiers moved the books of the French national library. Nowadays, civic action programs in rural areas have become a recognized role of military forces, and many basic technical skills have been conveyed in such a manner.

But the military institution as a stimulus to technology is found in its most significant form in the research and development complex. The heavy proportion of research and development funds concentrated in military, space, and nuclear energy has been noted. The latter two fields are sometimes disassociated from defense, but this distinction is quibbling. Often space is held to be nonmilitary, but its inherent nature has military implications. Nuclear energy was born as a weapon, and still is provided research priority because in spite of hopeful détentes in an age of mutual terror, its defense role is still paramount. Thus, it is not incorrect to say that the defense-related research provides the technological civilian spin-off which is the principal source of accelerated technological change in the modern world.

This comment is often made in a pejorative, derogatory sense, and the hope is expressed that all these valuable resources might be turned directly to civilian research. This sort of reasoning, however desirable, is simply not realistic under present conditions, and even if détente should become the world's way of life for the next century, it is unlikely that legislators with preaffluent-society ideas of economy would vote the expenditures for peacetime research. In the meantime it is important to work with the world as it is, and is likely to be in the next twenty years, which means that the engine of defense-related research is going to continue to generate massive technological innovations at about the same rate. It is true that the growth of expenditure has slowed down markedly recently, and it is only reasonable to expect diminishing returns in results, but such will not eliminate the requirement of adjustment to change.[2] New technologies and products would continue to flow, even if all funding of new research ceased, and the problem of transfers and adjustments would still be crucial. However, this is not to deny that research programs in fields other than defense will be necessary. The civilian spin-off from defense industries may not occur in places where needed, but this is another problem. The point stressed here is that military research-derived innovations are still at the heart of present-day change in the world's technological circuit, and will continue so indefinitely.

The nature of technological spin-off from defense research to civilian use is complex, and it is possible only to describe some of the obvious effects. It is probably safe to say that virtually all military research and development programs have some spillover effects. These may occur with a time lag, and it may be impossible to identify their incidence later on, but without the stimulus of military research and development, the remarkable advances in such industries as aerospace, aviation, and electronics would not have taken place. Actually, there is an entwining of civilian and military techno-economic activity, with much of the vast complex of the latter having some major components of the former. Civilian industry would not take the technological risk of investment on the scale necessary to obtain results if these researches were not required by contracts. Military contracts can lead to the foundation of entire industries; they may create new products or materials which would never have been developed, and they may create whole systems or ways of doing things which have been devised to meet military needs.

The computer industry may be taken as a prime example. Probably there is nothing which has affected modern industry more than the computer, but its origin can be traced to a World War II defense contract with the engineering school of the University of Pennsylvania, which produced the famous ENIAC. The applications of this device in industry and in administration and research have continued to snowball since that time, until people now refer to the "age of the computer." The gains in productivity for the United States and the world have been incalculable from this investment, which by orthodox thinking might have been termed unproductive military spending.

Many common civilian products would not exist without military research and development. The famous "TV dinner" of today has its antecedent in Air Force research which developed the idea of precooked frozen dinners for its aircraft crews on long flights. The idea was quickly taken up by commercial airlines, from which it has spread into general use.

Not only end products, but also the components of many modern facilities were derived from military sponsorship. The commercial airlines have been able to take over almost all of the aircraft engines and flight equipment developed by Air Force research. Transfers occur too in the most startling shifts between industries, with time lags, and even between different nations. For example, the famed Japanese high-speed train running between Tokyo and Osaka operates on ball bearings adapted from the United States F-104 airplane engine, which the Japanese learned to build under contract with American developers for United States military assistance.[3]

In the materials field the reach into space has forced the development of new metals and reexamination of assumptions in all technologies connected even remotely with the flight into space. John Welles has estimated that there are about 3,000 new products and processes which have spilled over from the national Aerospace Agency (NASA) into the economy.[4]

Entire systems or methodologies have come in as a result of military research and development. The whole field of industrial quality control has deep roots in military requirements, as has quality control's recent extension, the "zero defects" program. Methods of research and development themselves may have important value for civilian practice. For example, the transfer of fatigue testing methods for airplanes from the U.S. Navy to the Japanese is important for a country which wishes to extend the life of its airplanes.[5] But perhaps the development of new managerial techniques under the generic term of *systems engineering* has had or will have profound effects. The American Telephone and Telegraph Company has applied the idea to the telephone system, and some of the famed efficiency of that company is attributable to this approach.[6] The systems analysis approach of organizing team effort to solve problems on a multidisciplinary, interindustry approach is today being applied to critical problems in such social fields as housing, urban renewal, and mass transportation. Even the intransigent problems of the slums may yield to this powerful technique. Indeed, the cooperative, equalitarian, team approach as against the individualistic, competitive way may signal a major shift in practice throughout the economy.

THE MILITARY VEHICLE OF INTERNATIONAL TECHNOLOGICAL TRANSFER

Impressive as is the catalytic role of the military institution in evoking the new technology, it should not eclipse its equally important function as a channel of technological transfer. Military organization functions in this role within the home country, but abroad it plays an important part in the international transfer of technology from country to country. It has particular importance for developing economies, but even developed nations receive enormous benefits from its activities. An established channel ready at hand everywhere, it deserves careful study in a world irritated by technological inferiorities, and seeking ways of catching up with the leaders.[7]

The transfer of technology through military channels occurs in at least six major ways. These include demonstration, civic action, procurements, training programs, military assistance to foreign military establishments, and research exchange programs. These channels may include direct effects of the U.S. military presence abroad (or more generally any higher technical level military presence in a simpler technical environment). Alternatively, the transfer of technology may be quite indirect, operating through foreign military establishments working in conjunction with the United States military. The indirect impact may well be more important quantitatively in spreading technology to lesser developed countries. For the developed countries, the direct route, utilizing complex intergovernmental agreements and interfirm licensing arrangements, is more important.

The demonstration effect abroad is similar to the demonstration effect at home which upgrades the conscript by giving him new concepts and larger horizons. Abroad, a foreign military presence of a more advanced technical civilization demonstrates to the local people concepts which they may perceive as more effective ways of coping with the environment. The adoption of American products, practices, and methods often begins when the local people see and examine the military's example. This demonstration effect also occurs when foreign military trainees are brought to United States bases, ships, or stations or sent to the United States for training. On a higher level, it occurs with the outpouring of scientific and technical literature, some of which is classified, between the United States military establishment and friendly countries. The demonstration effect is an important first step in transfer of technology because it gives people a vision of what is possible and a motivation to close gaps which appear to them as technological.

Local procurement by United States military authorities can have important effects in upgrading a nation's industrial capability. Thus, contracts obtained by the Japanese in support of the Korean War effort were vital in laying the groundwork for the industrial rebirth of Japan. The American military stimulus to Japanese industry is widely recognized, but often overlooked is the vast injection of modern technology which occurred during that period. The Japanese were required to produce American items to American standards and specifications. Many of the items and production techniques were completely new to the Japanese. Under military procurement tutelage, they were able to receive invaluable person-to-person instruction in modern products and methods of manufacture. The severe rejection rates, applied as in the United States itself, provided the ultimate spur to mastery of the best modern practice. Entire systems of a techno-economic nature such as assembly-line practices, industrial control engineering, automated accounting systems, care and preservation, packaging, and crating were transferred at this time under military or quasi-military auspices. These laid the foundations for later industrial practice which has made possible the Japanese leap into a competitive position in modern world markets. In the original process, Japanese firms were able to contact American firms and obtain valuable license and patent privileges, setting the stage for long-term commercial collaboration.

Possibly most important for Japan, a country previously famed for low-quality merchandise, was that the Japanese were forced to learn to maintain standards and produce quality products to the specifications of the United States armed forces. It is impossible to describe how widespread this effect was, but if a basic area like food production and distribution is examined, one can grasp some of this diffuse but essential upgrading for modern life. In response to American military imperatives, the Japanese learned how to handle food under modern technological conditions. Old practices such as "hauling a carcass on a truck full of flies" gave way to refrigerated trucks. Night soil was abolished, and food handling practices were rigidly controlled. Food shipments took place in sanitized containers. These practices spread from the military procurement into the economy, and laid the basis for the modern tourist industry of today. Tourist stomach trouble, so common in Asia, is a thing of the past in modern Japan.

The impact of quality control is invaluable to the civilian market in improving the economic welfare of the people, but its benefit is also important in obtaining a better balance of payments position. Quality control is important for exports, making them acceptable to international standards, which is essential for a country trying to make its way in the world economy. Even small countries which have obtained procurement contracts have experienced this technical upgrading effect. The phenomenon takes place in peacetime at one remove through local armed forces procurement under advisement of United States missions, but such military action as Vietnam intensifies the effect. Military procurement in Taiwan, Hong Kong, Korea, and other Pacific countries has had much the same quality of upgrading effect as occurred in Japan a generation ago in the Korean War.

Military training is another important channel of transfer of technology to a recipient country. A vast technical transfer occurs by training local employees on United States bases and stations within a country, but the big impact comes through military assistance training programs which send people to the United States. From 1950 through June 30, 1965, the United States provided military assistance to eighty nations at an expenditure of $32 billion. Some 250,000 persons were trained in the program in the United States.[8] Many of these are key people who, after finishing their military careers, become managers and executives of vital firms or projects in their countries. Thousands are at the working level of the military logistics infrastructure or civilian industry holding responsible jobs made possible by this training. Still more persons became instructors and multiplied their technological impact by teaching in the armed forces' service training schools in the home country. In some countries, virtually the only source of electronics technicians and other skilled men is derived from veterans who went through these courses.

Of special interest is the worldwide English language program conducted by the Defense Language Institute of the Armed Forces. This specially devised system, which has been in existence more than thirteen years, has trained great numbers of people in the English language, and given them at the same time insightful instruction in the working of the United States system of government, economy, and social and cultural life. It has pioneered new techniques of rapid, systematic instructional methods which are universally studied and imitated. The system trained an estimated 100,000 foreign students a year at more than 216 sites in 41 countries.[9] In terms more human than the statistics, this system has experimented with training illiterates to become jet engine mechanics. In particular, the story is told of an illiterate shepherd in the Turkish army who mastered English without learning his own language, and went on to become a master jet mechanic. Having left the military, he is now alleged to be servicing international flights through Istanbul.

Military assistance programs, in addition to providing a worldwide technical program in a vast number of subjects, have built technical capabilities in many countries, transferring vast amounts of technology through programs in repair, overhaul, rebuilding, assembly, and manufacture of highly technical equipment in local facilities. These programs formerly were restricted to developed countries in Europe and to Japan, but now self-sufficiency programs are emerging for small countries. Under these programs, American defense contractors make available under license the rights to service, rebuild, and manufacture the advanced equipment originally researched, developed, and produced for the United States military. In addition to making available technical information, data, designs, and specifications, and providing substantial amounts of the funding, the American managers, engineers, and technicians who did the original work also are made available to the cooperating nations at locations abroad. Conversely, the foreign engineers, managers, and technicians are brought to the United States factories. Bilateral programs and consortia in Europe built advanced aircraft and other weapon systems involving billions of dollars. In Japan, Mitsubishi and Kawasaki Aircraft learned the most advanced production skills from Lockheed and North American companies. In electronics, one technical journal in the field maintains that seventeen years of electronics lag in Europe was compressed in these military contracts. Tropospheric scatter systems of communications and advanced electronic systems of controlling aircraft like BADGE have been transferred or are in the process of transfer.[10] In Japan's case, popular wisdom has it that there is but a six-month lag between the fruit of research being made available in the United States and its possession by the Japanese. In Europe's case, the lag may be longer, but if so, the reason lies in the dislike of imitative practices which borrowing this technology entails.

Developed countries prefer to be involved in the creative research process itself, and they are not impressed by transfer of an ongoing system, perhaps being less anxious for commercial applications. In any event, even research activity itself is being transferred via the military channel. There are programs such as the Defense Data Exchange agreements in which specific research systems are made available by the United States in exchange for such systems abroad. Cooperative research and development programs between the United States and friendly countries have begun to be developed in which participating countries may take part of a research program and thereby gain access to the whole. International military exchanges have been established under which American and foreign scientists and engineers are able to work in each other's laboratories for periods up to a year. Lastly, a vast amount of technical literature is sent overseas every year. The United States Department of Defense and the Clearing House for Federal Scientific and Technical Information make available every year literally millions of documents on research and technology. These latter channels, however, should not be overemphasized because they are not buttressed by the person-to-person, equipment-supported package which effective transfer of technical know-how seems to require, but they must be counted in the total military transfer of technology.

IMPROVING THE MILITARY CHANNEL OF TRANSFER

It would appear obvious from the foregoing that the military institution is a prime channel for improving the technological conditions of countries, both developed and developing, which are lagging in technology. That this channel is less recognized than it should be stems from the grudging credit which the public at large accords to the military institution. Moreover, the military itself tends to underpublicize its activities because of security classification requirements and other reasons which contribute to the lack of awareness of the valuable technological role which military organizations play in disseminating technology. For example, military commanders concerned with their primary mission are reluctant to become involved with such secondary objectives. But lack of general recognition should not be permitted to obscure the military's actual and potential contribution in closing technological gaps. As with the commercial channel, the vitality of the military lies in providing a package of dynamic, reinforced technology to the recipient countries. Demonstration, information, drawings, specifications, and even person-to-person contacts are not adequate when taken individually, but require reinforcement in the composite "package." Much of this holistic quality already exists in military technological transfer; improvement would appear to lie in strengthening such tendencies.

However, there are some preconditions which should be noted, one of which is the need for greater publicity. Publicizing its technological activities to secure greater recognition and acceptance of its role would constitute an important antecedent step. The negative image under which the military organization labors should not be permitted to generate continued silent, secretive responses. Classified material has a role, but usually it far outlives its usefulness. Another part of the problem also requires that the military organization accept a changing role for itself in a changing world. Professional military men must accept responsibility for a larger role in world technological transfer. To some extent this has happened, as witness the Air Force's "Foreign Technical Officer" (FTO) attaché abroad, but his emphasis is usually on watching foreign developments rather than taking on creative tasks. A more imaginative institution-building approach is important.

What could be done in each country depends on the country's technical level and situation and on the country's propensity to accept and adjust to new technology. Many factors are involved in each country's case, but some general policies could be developed both in the United States and other countries for strengthening the technological transfer role of the military.

For underdeveloped countries the structuring role of the newly developed armed forces can be made into a still more vital force in order to upgrade technological levels. American advisers have trained and procured with an eye largely focused on creating an efficient military organization. The civilian impact has been treated as a by-product, usually unrecorded and unhailed. Assuming that improving technological transfer is desirable policy, and most men of good will hold that it is, a more careful record of what happens in an economy as a result of military inputs is a first step to wisdom. When the record is established, action to improve it may become obvious.

For one, we need to know how each man who has received training fits into the civilian economy after his discharge and how his military training relates to his later civilian occupation. It is a good hypothesis that there is a tremendous influence. One hears remarkable stories like that of the illiterate Turkish shepherd turned jet mechanic mentioned previously, but verification, analysis, and, even more, in-depth examination of many cases is required.

For "package" technology, recent policies to encourage local repair, overhaul, and rebuilding facilities for United States equipment in simpler economies is likely to provide a greater thrust for development than the military has yet sponsored. The option to build local capabilities involves other benefits and costs to the United States and the developing country, but effective technological transfer of ongoing systems is a consideration which is not to be dismissed lightly.

Procurements are a particularly strategic level for upgrading local facilities. It is true that procurement officers must seek the most efficient and immediate source of supplies, particularly when faced with crisis situations, but where possible, it is established procurement policy to seek out alternative sources of supply. Small, less developed countries offer alternatives, but they must be encouraged to do so. A Korea or Taiwan is obviously overshadowed by a Japan, but it is not desirable to concentrate the contracts in one source. When time is not essential, the less experienced country's industry can be spoon-fed to do the work. In time, it will acquire expertise and speed.

For the developed country it has long been established practice to participate in the military activity connected with such alliances as NATO. The arrangements include information exchanges, sales of equipment, cooperative production schemes, and/or cooperative research and development activity. These conditions are often virtually unknown to the public at the time they take place because of security reasons. Even years later, little publicity is released. In this way a great deal of the military channel's amazing effort is lost to public view.

Within the framework of military security, more should be done to publicize the beneficial aspects of military transfer of highly advanced technology. First, there has been an enormous transfer of technical information from the United States to other friendly, advanced countries. Over four million technical documents per year flow to Western European countries. Enormous amounts of technical data in the form of manuals, drawings, and specific information accompany sales, loans, and co-production or co-research schemes. Obviously, a simple way to increase transfer of such information is simply to increase the amount provided and to expedite security declassification in order to make this material more generally available to the scientific and technical community in advanced countries, or for that matter, in any country—even the socialist world. However, there is a limitation on how effective technical reports can be in the absence of equipment and personnel. As previously stressed, technology is part of a package of other inputs, and its effectiveness in textbook isolation is limited.

More effective are sales of equipment accompanied by expertise. Probably the more comprehensive the system, the more effective the transfer. Yet while sales of equipment or systems upgrade the recipient's technical capability through operation, maintenance, and even rebuilding activity, it must be recognized that what is a blessing to an underdeveloped country, or even a semi-developed country, may be somewhat galling to the people of an advanced economy. The purchase and use of American equipment makes the country to some degree dependent upon the United States. That this is an efficient business transaction does not weigh as much on the scale as technological nationalism. The huge investment and risk of United States effort are forgotten at the moment of delivery of a finished system which has been planned five or more years before. Some part of this problem is the fault of the advanced country which is borrowing the technology for not having had the foresight to appropriate funds previously in order to bring out the product. The British failure to develop airplanes like the F-111 has placed them in need of buying them. Lockheed's announced sales of CL-879 airplanes having properties of both a helicopter and a fixed-wing airliner, alleged to incorporate properties of the Fairey Rotodyne project abandoned by the British government in 1962, makes bitter headlines some years later.[11]

The way to improve this situation is for the advanced country to commit more resources to research and production efforts, or somehow to make existing commitments more effective in terms of finished products. Alternatively, there is the possibility of the country joining the United States in more co-research and co-production programs. It is possible for the United States military establishment to encourage programs of joint cooperation in research and production beyond the many technologically advanced systems which have already been shared such as the F-86, F-104 aircraft, Hawk, Sidewinder, and Bullpup missiles, and shared space contracts. But it must be recognized that these sharing activities usually will not be financially profitable for the cooperative country's industry because of the negotiated character and cost-sharing nature of these contracts—practices which are largely not in use abroad.

Lack of sales or markets will also limit military or commercial exploitations and profit opportunities. The commercial spillover will be a function of the foreign management's effort in applying the technology learned to other new products for military and civilian usage. Nothing will relieve the cooperating country—both its government and private sector—from the burden of their own sustained effort in the world's technological race. Yet more cooperation along the lines of the proposed Defense Common Market offered by the United States would help them. Basically, it would be to the advanced countries' advantage to continue to accept the U.S. Department of Defense's military assistance sales program. Much civilian technology transfer occurs through military sales. A good deal of this kind of thing would occur in a Defense Common Market, which is actually in small scale operation.[12]

Exchanges of engineers and scientists under United States military auspices have been developed and are a fruitful field for further expansion. Under this program there are international technical committees which are established to exchange information in individual technical areas. In most cases these exchange programs provide a concrete form of technological transfer, but they lack the project package orientation. Intensification of these exchanges would appear to offer scope for improvement. Even more effective, at least for the individual, are exchange programs between United States military research organizations and counterparts in other countries. A European scientist or engineer who spends a year working in a United States defense laboratory and his American counterpart who goes abroad, offer many opportunities for personal professional broadening and cross-fertilization of ideas and techniques. But valuable as these individual activities are, they do not transfer technology in the package capability sense which is necessary to produce final products in a competitive world.

It is this transfer of capability which frequently appears abroad as derided dependence. As an alternative, nothing will substitute for a nation's own effort, its commitment of resources, and its determination to carry through projects to completion even at great risk of sacrifice. Perhaps some middle ground of cooperation is possible if developed countries are a little less sensitive about insisting on their own developments. American military planners, given the willingness of United States industry to cooperate, could devise programs of still greater cooperation. One that has been suggested is to establish preserves expressly for foreign contractual participation in research, development, and production. On the same principle that contracts are "set aside" for small business in the United States, some part of the contracts may be reserved for bids from the industries of friendly countries. Such a program would not be a substitute for the foreign country's own efforts, but it would be a way of gearing developed country industry more effectively into the magic of military research and technological transfer.

It must be stressed that this type of scheme or any other would depend heavily on the willingness of American industry and the United States Congress to cooperate along such lines. As will be discussed in the following chapter, American industry is taking on multinational characteristics rapidly, but it is still United States-based, and it directs its activities largely from a United States headquarters and viewpoint. Yet for various reasons, it may find the role of subcontracting research, development, and production contracts less unattractive than might at first be supposed. Similarly, opposition from Congress may and has occurred. There is strong pressure in Congress to drop Project Mallard, a joint program with Britain, Canada, and Australia to develop a tactical or automated communications system, a billion-dollar undertaking involving numerous American and foreign companies. Such efforts are inherently "turbulent and trouble-ridden" according to the House Appropriations Committee.[13]

NOTES TO CHAPTER 9

1. B. A. Battison, "Eli Whitney and the Milling Machine," *Smithsonian Journal of History,* I (Summer, 1966), pp. 9–34.

2. For the slowdown in the rate of advance of United States Government Research Expenditure, see National Science Foundation, *Reviews of Data on Science Resources,* No. 17 (February, 1969), NSF 69–12, p. 2.

3. See Daniel L. Spencer, *Military Transfer of Technology* (Washington, D.C.: United States Air Force Office of Scientific Research, 1967), Technical Report No. 3, AD–660–537, pp. 94–95.

4. John G. Welles, *et al., The Commercial Application of Missile and Space Technology,* Denver, Colo.: University of Denver, Denver Research Institute, September, 1963.

5. Daniel L. Spencer, *Military Transfer of Technology, op. cit.,* p. 101.

6. J. A. Morton, "Model of the Innovative Process," *Technology Transfer and Innovation, Proceedings of a Conference* (Washington, D.C.: National Science Foundation, 1966), pp. 21–31.

7. For an introductory survey based on the United States Military presence in Japan, Korea, and Taiwan, see Daniel L. Spencer, *Military Transfer of Technology, op. cit.*

8. United States Department of Defense, *Military Assistance Facts* (Washington, D.C., 1966), p. 31.

9. Defense Language Institute, *English Language Training Program* (Washington, D.C., 1964), p. 1.

10. Details of some of these agreements are available in Daniel L. Spencer, *Military Transfer of Technology, op. cit.*

11. "U.S. Offers British European Airways Super-Plane We Scrapped," *Sunday Express* (London), April 2, 1967, p. 1.

12. J. L. Trainor, "Can U.S. Maintain the Momentum of Its Military Export Sales?" *Armed Forces Management,* January, 1967, pp. 36–40.

13. *Business Week,* December 27, 1969, p. 26.

10
INTERNATIONAL BUSINESS TRANSFERS

If we accept the view that the best way to transfer technology is through a package, the channel of international business is by all counts the most effective. But it is also a channel fraught with controversy. There is an inherent conflict between national states which object to nationally unintegrated entities controlled from New York or foreign centers directing a subsidiary or partner within their domain. The old cry of economic imperialism is raised even in the largest and most friendly countries. Any realistic assessment of the international corporation as a channel of technology transfer must balance its efficiency in transferring technology against the fears of domination which it evokes abroad.

There is little doubt that the receiving country gains far more in modernization than it loses in economic independence, but control over technology is important to the people of a nation and its government. Improving the efficient business channel of transfer means largely devising techniques of coping with the inherent conflict between the national state and the international corporation. Particularly controversial for developed countries is the independence of the technological research base. A crucial question is how innovation can be rapidly transferred across frontiers through the business channel without causing excessive tension or impinging on national sensitivities.

THE EVOLUTION OF TRANSNATIONAL BUSINESS AND TECHNOLOGY

There is a long history of firms doing business in other countries, as witness Renaissance Italian trading firms and banks in North Europe. The nineteenth century saw British, French, and other national companies investing in and building industries all over the world including the United States. Before World War II, the United States had some foreign investments, notably oil and other extractive interests which tended to be concentrated in Latin America and Canada, but such investment momentum as was achieved was cut short by depression and war. Moreover, the colonial world was dimensionally different from the world today, although precedents generated may have persisted in some cases. For the most part, the surge in investment following World War II may be said to have started afresh because it was based to a large degree on the new technology. Unlike the previous activity, the investment was directed toward the advanced countries of Europe, though as time has gone on, the entire non-Soviet world has been considered, and even socialist states like Yugoslavia have invited foreign capital to come in[1] with Western patents, processes, sales contracts, and simple managerial know-how.

Motivated by search for profit opportunities, American business saw the national incomes of Europe rising at a more rapid rate than in the sluggish American economy of the 1950's. The Europeans were attracted by the new consumer products and by the advanced technologies which made them possible. Another big impetus came with the threat of exclusion of American products by the formation of the European Common Market. Once a firm was inside the Market, it could manufacture and sell in a free-trade area comparable to that of the United States. Exports in the first period after the war had been easy for American industry, but as European industry caught up through imitation and innovation of its own and as the threat of an external tariff loomed, it was good business to locate a subsidiary or affiliate behind the tariff walls.

There were other advantages. The importance of local relationships with customers, nearness, and the importance of after-sales care also played a role. Then, the product itself had to be tailored to meet local requirements. On the cost-of-production side, in spite of productivity and inflationary offsets, one could get the advantages of a cheaper labor in manufacturing, and the products, despite the protests of American labor, might be imported back to the United States. For example, in electronics Japan assumed a role of component parts supplier for well-known brand American products, and Hong Kong is following Japan's lead. In Japan's case another advantage lay in turning over slightly dated machinery and equipment which was very modern to the Japanese but a vintage behind the latest models in the United States.

All of this is straightforward practice of the business institution. Its novelty for Americans was that business crossed international frontiers, and managements had to learn to get along in a new, more difficult environment. With exceptions like International Business Machines, Singer Sewing Machines, National Cash Register, oil companies, and some manufacturers with foreign assembly plants, American business had little experience in the international field, and much of what they had was concentrated in Canada or Latin America. But management learned, and in recent years the idea of having an international division or setting up a product structure which takes account of the foreign factories has become an important part of American business practices. Indeed, it has become something of a fetish to "go international," and many companies have acquired foreign connections as they would such other trappings of modern business as ornate offices and management consultants.

But for the serious international company with experience and maturity in the international arena has come recognition of the difficulties of doing business in the foreign environment. The opening of dialogues with foreign businessmen and governments has brought recognition of their viewpoints, and effective communication has resulted in change of attitudes. Businessmen have been forced to make many intellectual and actual adjustments in their thinking. They are beginning to see that the simpler ethic of old-fashioned operations must have a broader social base. International business is rapidly becoming a socially conscious component of the industrial West. Thus, most companies try to bring foreign people into the corporation's family, as implied by the overused term *multinational corporation*.

Some truly multinational corporations may already exist, and it may be that many will evolve in the future, but the fact is that for most international companies, control from the head office in the United States is still the fact of life which everyone abroad knows. There are, to be sure, some large European-based corporations like

Royal Dutch Shell, Unilever, Philips in the Netherlands, and Nestles of Switzerland which have been in international business for a long time, but though these seem less conspicuous than American companies, their controls are not necessarily different. Yet they do not seem to arouse the same antagonisms as does the American-based company. They appear to be more multinational because their volume is spread among many countries, whereas with some exceptions like International Telephone and Telegraph and National Cash Register, the volume of international business for most American companies is relatively marginal to their domestic sales.

Widespread or not, the existence of control vested in the United States has nettled European and other countries. The large volume and seemingly ever-increasing size and scope have threatening overtones to foreign countries. A few companies could be absorbed easily, and in fact have been in the past. As in the tax field, in which it is the new or additional tax which causes the outcry, the impact of the established foreign company which has done business for a long time provokes little comment. When the new company or the additional investment seeks to come in, the term *invasion* is heard.

As they have become increasingly sophisticated in the international field, American managers have indeed been moving toward a global point of view, but the word *global* has a special meaning. They recognize that there has been an increasing economic unification of the trading world as communications have shortened time, and trade and investment barriers have declined. Their global vision is to match an order for low-cost products from, say, Nigeria with low-cost manufactures from the Taiwan plant—all done instantly by computers in the New York office. There are, however, problems obstructing this vision. Within the company its American plants find foreign competition unattractive, American organized labor doesn't approve of exporting jobs, and the United States government is still more uncomfortable with immediate balance of payments difficulties. On the foreign country's side, businesses and governments do not like the idea of being mere recipients of technology and order takers.

Actually, there is a kind of love-hate relationship in attitudes toward American business in advanced countries. The countries want the modern products and the technology, but they dislike the price which they feel they must pay. There is clear recognition of the efficiency of American plants and equipment in local industry, but there are fears of economic domination. The French attempt to exclude American business is the extreme example, but in all countries people dislike losing the decision-making power in a large proportion of a major industry. The fear is justified to some extent in that a parent company with interests in many countries does not have the same interests as those of the local plant. As the systems analyst puts it, the nation's desire to suboptimize (say, exports) may not coincide with the optimums of the global system. Research and development effort is a particular point of issue. The international company typically centralizes research in one country, usually the United States, because of efficiency considerations and because of military-related contracts. On the other side, the people of developed countries want to take part in the advance of technology and science; they resent a mere production role for their national industry.

The American global managers are aware of these feelings, but they are often unable to reconcile their needs for worldwide administrative organization with the local desires. The problem remains largely unresolved, and in fact it has been intensified by the new international emphasis of American domestic companies which formerly had little international interest. In earlier days when a company had overseas branches, they tended to run themselves fairly independently. Today, the vision of the "multinational corporation" has made for more centralization and global planning to the detriment of such previous overseas independence as existed.

In the developing countries the problem is not a desire for independent research or scientific traditions but is derivative of the heritage of colonialism and the Marxist refrain of exploitation. To the Western company, nationalization is a real threat as are ill-conceived legislation, unreconstructed bureaucrats, exchange restrictions, and many other hazards to foreign investment. For a long time these deterrents have blocked much overseas investments in developing countries, except in such extractive industries as oil. This seems to be changing now. The American companies' new international outlook has brought a new willingness to search for opportunities even in hostile places in order to attain a more complete world coverage.

There is also a willingness to work with governments in mixed arrangements which would have been inconceivable a few years ago.[2] Not all companies are so venturesome, but going into backward, truculent countries is roughly comparable to domestic developments in product diversification, which involve branching into new lines and establishing beachheads in other industries in which a company has no experience.

The principal difference in the foreign environment of an underdeveloped country is that the outside company must exercise much care in assuring local people that they will benefit from the investment. A developed country may be able to see a community of interest and long-run, indirect benefit accruing via a circuitous route. The developing country has more naiveté and immediacy of vision. Its people dislike the isolated enclaves of foreign interests whose day is rapidly drawing to a close. The suspicion and socialistic dogmas have to be countered by positive steps to integrate company activities into the society so that the alien origin is blurred or even lost. International business generally recognizes that it must take on much of the interest of the country concerned. Yet many international companies have not been willing to pay the price; others have paid only after hard bargaining.

On the other side, the socialist-included governments have begun to see that their bureaucratic regulations and other unfavorable conditions have deprived them of the hard-driving international companies. The obvious failure of grandiose plans and government enterprises to show results has been driving developing countries to a more realistic rapprochement with international companies. The rigid positions of a few years ago have eased somewhat. The dialogue which has been going on now for some time has generated some changes in attitude on both sides.

The advantage to the developing country is that it can secure the vital technology which it needs from the going industrial concern. Capital, skilled and efficient management, and foreign, capable engineers come together with the designs, blueprints, instructions, and technical data of all kinds. Technical difficulties can be straightened out by long-distance telephone to the home plant. Long-range modern planning at the company level, jobs for the developing country engineers, enlightened labor relations, ready-made marketing outlets for exports, skillful cultivation of home demand—all these become possible if the foreign complex is accepted. And acceptance means the high standards, high profits, and at least some international control which accompany the foreign firm.

ALTERNATIVES FOR BUSINESS TECHNOLOGICAL TRANSFER

Against this background, what are the alternatives open for closing gaps and transferring technologies more efficiently through the business channel? In theory, it should be possible to treat the world technological system as just that—a system. If one asks how best to transfer technology from country to country, systems analysts who have been so successful in applying their computer capabilities for analyzing and solving complex problems of space and weaponry may be able to apply their talents to this problem of a rational and equitable world technological system. Such attempted applications are in the offing for such social problems as urban renewal or poverty, and the world business and technological system may be susceptible to similar treatment. If such a system could be designed, it would have to take into consideration the kinds of competing objectives which are implicit in the problem. Alternatives for the several parties involve conflicting courses of action, which may be, at least on first analysis, incompatible. Yet it is always the hope that where dialogues take place among rational men, trade-offs are possible and some effective *modus operandi* can be worked out. A review of options facing some of the broad divisions of interested parties may be of some help in a move toward a more effective world technological system, in which the efficient bearers of effective package technology—the international business companies—are encouraged with adequate incentives, and recipients are assured that accommodation to the changes does not mean disintegration or demoralization.

Excluding such possibilities as the International Technology Authority, outlined in the previous chapters on military and government channels for business and government in the developed countries, there are three possibilities for foreign companies: (1) meet the American competition, (2) join the American competition, or (3) join together to meet the American competition. The inflow of United States-based, international companies with military-related, research-derived technology is in the nature of a challenge-and-response situation which, uncomfortable as it may be, forces change. Meeting the competition is the first alternative, which usually means becoming more efficient, though a strategic government contract may also be the lever of survival. For efficiency, in the conventional sense, may also require alteration in established ways of operating. It may mean thinking big: merging smaller companies together and seeking larger markets. It may mean hiring consultants, even American consultants—the adoption of American methods—to fight fire with fire. It may mean giving up much leisure for managerial elites who have patterned ways of life. With such adaptations and with ruffled tempers, the new technologies and systems get transferred, and the world's technological system starts generating a new problem. Thus, the old order changeth.

Alternatively, there is the realistic course of joining the American invaders. At the industry level, this may mean merging smaller and inefficient companies into the big international company. Englishmen, Germans, or Japanese find that it is not the end of the world to be controlled from the New York office, and the price of survival may not be as high as is believed. With such acquisitions the international firm becomes more what it claims to be—multinational—and the scope for individual achievement is open to able Englishmen, Germans, or Japanese as well as to Americans. Possibilities and opportunities are enlarged for every one, and this is after all one definition of free choice, a fundamental value for most of the world.

The third alternative for the advanced country's business is to join together with business in other advanced countries and create large entities which will make possible the development of independent bases of advanced technology. This creation of international companies as counterweight balances to the United States-based company parallels government moves toward common markets. In the latter connection, some of the same subordination of national pride must also occur, but the local people will not feel as dominated by foreign interests which will then be more equal in size to domestic companies. Instead, they will have to make heavy adjustments to the differences in methods of technique and management which obviously exist among advanced countries. In an alleged first of its kind, a French computer company and an English computer firm are recorded to have joined together after attempts by American computer interests to take them over. The merged company's services will be available throughout Western Europe and will embrace all facets of design and use of computer systems in business, defense, and scientific fields.[3] Some combination of these alternatives may also provide viable answers for government and businesses of the developed countries.

Some of the foreign companies are as efficient and well managed as any in the United States. Indeed, the United States itself has its share of backward managements and recalcitrant labor as such lagging industries as shipbuilding or railroads suggest. Much of the problem is not in company hands but results from the decisions of government civil servants and policy people to expand and contract programs and to commit (or fail to commit) resources to backing promising industries and research programs. However, while this is changing, many advanced country businesses still suffer from paleolithic managements and engineers concerned with status, seniority, family, deference, and other liabilities which will have to be jettisoned to produce competitive and viable responses to American incursions. As an antidote, the motto of Japanese businessmen quoted by an admiring British industrialist recently returned from Japan is instructive advice: "Find out the best practice in the world and improve on that." [4]

OPTIONS FOR AMERICAN BUSINESS

On the side of the dynamic American business in the international field before the imposition of balance of payments restrictions, the alternatives were: (1) to keep up the drive into the developed country territory, (2) to slack off and let up the pressure, and (3) to keep pushing but absorb the opposition with greater representation and opportunity for participating creatively in the modern dynamic technology. The first alternative offered more of the same advantage as the initial overseas movement. Basically, the American firm has some monopolistic characteristics through its new research-derived technology, its know-how, and its management which are not available to most of its foreign competitors. Added to this, it usually has greater financial strength through its size because of the large scale of the American market. Further, the American company often has been greatly stimulated by huge government spending on military-oriented research, development, and production. By going into other developed countries, higher profit ratios and high growth rates are possible for relatively small incremental investments. The idea of internationalization is associated with diversification—taking on new product lines and entering new industries in the home market. This is part of a general tendency of American companies to hedge or protect their position in a constantly changing world. Opting for a continued drive in buying up or taking up good investment opportunities in developed countries is consonant with a basic appetite for greater diversification of operations and of course larger profit opportunities.

American business is still relatively new at international affairs. It still has many issues and problems to work out, and it is still evolving. The alternative of continuing its drive is probably the one most attractive to the international companies because of their inherent dynamic. The penalties, however, for this choice are likely to be increasing enmity and frustrations from press and people, increased government control of investment, and possible nationalization of business.

The opposite course of slacking off has been imposed, at least to some extent, by United States government restrictions on further investment in advanced countries for balance of payments reasons. However, alternative sources of financing in Europe may be used by American companies even at greater cost. Therefore, it is by no means certain that the restrictions will stop further investment. It is true that the growth curve of American investments may be reaching a plateau stage with the advent of diminishing returns. Even before the imposition of balance of payments restrictions, the big-profit, easy-money days for American firms were probably finished. Yet the momentum of overseas investment has been very great in the past, and continuation of the drive cannot be ruled out.

In the future, some middle ground of continuing to expand into advanced countries, but with more attention to the opposition, would appear to be likely and to yield maximum advantages. If American corporate management wants to claim multinational companies, the other nationals have to be given a share in control or at least have more opportunities for participation, particularly in research and development in local companies and plants. Much has been done, for example, by way of training schemes which bring people to the United States, but more seems to be required. Yet many big companies like General Motors insist on the complete or single ownership principle of its overseas subsidiaries and do not take kindly to sharing control. It is argued that only in this unitary way can a company meet competition and change in the world economy, as well as deal effectively with local nationalism. Such companies argue that unified ownership can respect national governments and national goals but that the companies must have control if they are to be effective as international entities.

In support of this position, one of the more adroitly conceived overseas operations which is 100 percent owned but heavily involved with and solicitous of the local culture and social relations is the new Sears, Roebuck and Company store in Barcelona, Spain.[5] The lengths and expense to which Sears' international executives have gone to reassure local interests are impressive. Such innovations as credit cards issued to the elites in the city and decor suited to the taste of the local people are among their methods, in addition to heavy commitments to undertake exports for the government. Perhaps most impressive is the catalytic effect on local department stores which are already restructuring their operations to meet the new competition. Following their achievements in Latin America, Sears would appear to be a model of the beneficent effects of a sophisticated, carefully planned American company's entry into a foreign market. Yet objections are always present. For example, local Barcelona manufacturers object strenuously to the Sears practice which excludes the manufacturers' names from Sears-sold merchandise.

Joint ventures with local partnership have merits and demerits which the literature on international business catalogs extensively. For present purposes, the point is that the acceptance of a local voice in controlling company destinies gives it a more domestic acceptability, though at the price of hampering international company policy and operations. In effect, this means that transfer of technology through business channels is slowed down. Resistances build up behind the local partner, and the process of adjustment which must take place is delayed and hampered. In contrast, the "hard-nosed" policy of single ownership installs the new technology, and adjustments are forced to follow. The ultimate risk of expulsion of the foreign company by government action is a calculated one. Generally, there is a trade-off in each situation between making the investment operate efficiently by American standards and the placating of local pride and retention of slow-moving local practices.

Another method of local participation is authorization for local subsidiaries to do more research or even to build research centers. There is a healthy respect for European and Japanese engineers and scientists in the United States. Britain particularly has brought out many ideas, such as the pivotal-wing, supersonic airplane, which were later developed in the United States. Decentralization of research effort and establishment of laboratories abroad will give local engineers and scientists more opportunity within their own countries, and counter the emigration movement to the United States.[6] Indeed, big companies seem to be aware of the need to conserve scientists in Europe for their European operations. It would appear to be the smaller companies and those with minimal interests abroad which are advertising to lure people away to the United States.

Still another way to provide local participation is to offer more opportunity for outlets of European company technology in American manufacturing. Some of this already exists, as witness third-generation jet engines from Rolls Royce installed in the Lockheed air bus, the jumbo airplane, but more could be done.[7] That is, European industry should get more of a share of contracts for development and production in advanced countries. This concept would tie in with "set-asides" of government contracts discussed earlier. Still more licensing to European industry will also be helpful in bridging gaps. In the past, American business has been following a generous sharing policy with respect to technical information which contrasts with secretive traditions in other advanced countries. Under this alternative, American business would develop a program for easier and wider transfer of technical information with or without government cooperation. By itself this would not be enough, but backed up with contract or subcontract opportunities, it would have positive ameliorative effects.

This last alternative for American-based international business, namely, continued expansion with careful attention to advanced country participation, will not ease all the difficulties which developed countries are experiencing in coming to terms with the new technology, but it lessens the image of international business as a devouring monster. It will also pave the way for cooperation among advanced business complexes in the United States and those in other developed countries to attack the problem of transfer of technology to the developing countries. It is here that the gap is most wide, and better relations among the advanced countries will lay a foundation for international business to play a bigger role in developing countries. The decision makers in under developed countries are faced with the alternatives of continuous stagnation in failing to come to terms with the dynamic ethic of Western business or an opportunity to obtain technology in its most effective form. Better relations among the developed countries may make it easier for recalcitrant elites in poor countries to choose sensible, cooperative courses of action.

Lastly, there is another excellent reason for cooperation rather than competition, namely, forestalling foreign advances in the home market. The challenge to the challenge of American business is already making itself felt in the inflow of European and Japanese imports to American market at home and abroad.[8] Investment by foreign business in the United States to counter the flow into Europe may be regarded as sensible countermoves in the chess game played by international oligopolists.[9] Large American companies are generally larger than their foreign counterparts by a factor of two or more, but this ratio has been declining. Merger movements abroad, for one, have been an important equalizing factor. If expansion with cooperative participation can forestall such counterattacks, it would seem a rational course of action. A forecast for the 1970's would predict it as a probable course of action.

NOTES TO CHAPTER 10

1. *Time Magazine* (Atlantic edition), April 7, 1967, p. 29; *Business Week,* December 20, 1969, p. 74. Also supra, Chapter 6.

2. For a discussion of the historic problems of government and private business participation, see D. L. Spencer, *India, Mixed Enterprise and Western Business* (The Hague: Martinus Nijhoff, 1959).

3. The firms are the British Computer Analysts and Programmes, Ltd. and the French Center d'Analyse et de Programmation. *New York Times* (International edition), April 7, 1967, p. 7.

4. "Two Island Nations: A Study in Contrasts," *U.S. News and World Report,* January 16, 1967, p. 59.

5. *Business Week,* April 15, 1967, p. 92; *Ibid.,* December 6, 1969, p. 208.

6. "Tapping Talent Overseas," *U.S. News and World Report,* February 26, 1968, p. 60.

7. The Lockheed Air Bus Agreement with Rolls Royce was announced March 28, 1968, and reported in *New York Times,* March 30, 1968, p. 1.

8. John B. Rhodes, "The American Challenge Challenged," *Harvard Business Review,* Vol. 47 (September–October, 1969), pp. 45–57.

9. Stephen Hymer and Robert Rowthorn, "Multinational Corporations and International Oligopoly: The Non-American Challenge," Yale University, Economic Growth Center, Discussion Paper No. 75, September, 1969 (mimeographed), pp. 1, 11, Tables 3–7. Also presented at the *American Economic Association Meeting,* New York, December, 1969.

11
GETTING ON WITH THE JOB

The international spread of technology is an increasingly important matter. In a less sophisticated world, it took care of itself, but in a world of telescoping time and instant information, more attention has to be paid to the mechanism of transfer. What is it and how does it function? How can we transfer the technology or participate in its benefits? What is the transfer mechanism, and how can it be improved? These are the questions explicitly framed or implicit in the minds of persons one or more steps removed from the central powerhouses of research and development which are generating the waves of innovations. Even those close to the process feel the need to explore its dimensions, as witness the Symposium on World Trade and Technology convened at the opening of the new National Bureau of Standards building or the Denver Research Institute's Snowmass-at-Aspen Conference on transfer of technology, which asked similar questions.[1]

WORLD TECHNOLOGY TRANSFER SYSTEM

There is a worldwide technological circuit which has emerged in the modern world. It begins with research in some technology which often is connected with military or related activities. New processes or new products quickly spill over into the civilian economy in the same or newly visualized usage. These transfers of technology take place through government, military, education and other channels of information and planned effort, but technology transfer is most effective in the market business sector where the incentives of corporate and private gain are operative. All channels are interrelated, but in the West, particularly in the United States, a system of transnational business has grown up in parallel with and intricately bound up with the new technology. It seeks to convert the findings of its research centers to sales and profits at home and abroad. It is manned and managed by men of talent who seek to maintain the position and growth of their companies in a very competitive market place. Perforce, they have created large aggregates which act as magnets for further talent at the center and loom as threats to establishments abroad when the technology is exported.

The technology is transferred abroad either by investment or license of some sort, sometimes in connection with military alliances. The transfer system is remarkably effective, and within a short time the innovation is in use even in remote corners of the world. After some time the technology or the technological system is absorbed, locally reproduced, and eventually improved upon. Industries with leadership in one country decline and cede their place to other up-and-coming industries in other countries. These latter may start out as uninspired copies, but through application of brains, energy, and good luck, they find themselves the leaders, even exporting to the country where the technology was originally developed.

The market equilibrium system has transferred technology in the past and still does. If you want technology, generally the rule is that you go to the market and buy it. In most cases it is for sale, and it is relatively cheap. But the sale comes in a big package, and the buyer has to be willing to pay the price, not only in terms of having the internationally acceptable currency to do so, but also in the sense of having to make the necessary adjustments to be able to use the technology after it has been purchased. In the past, sooner or later countries have paid the price, made the necessary adjustments, and gotten on with the job.

The trouble is that nowadays the whole process has been accelerated. The conscious effort involved in research and development to make discoveries, cultivate them, and eventually bring them out in actual practice has been perfected and multiplied by the commitment of huge resources to the process. Large business aggregates become larger through mergers and other sources of growth and threaten the less concentrated firms. Technological and trade gaps develop and seem to increase in size. Other considerations become entwined with the original business of getting the technology out and to work. Frustrations at the inability to mobilize resources on a similar scale cause resentment among businessmen and governments abroad. Another resentment-producing factor is the scale of required change which disrupts the traditional order and challenges fundamental attitudes and ways of conducting life. These changes might be made in due course, but the accelerated pace requires larger effort. Technological gaps loom large, and there is public pressure to correct them.

The international trading world in which technology moves freely, albeit for a price, is also a world of national states. If frustrations and resentments cumulate, nationalism is invoked. This may result in exclusion policies such as the French have tried, the cultivation of uneconomic local research efforts, or, in the final analysis, the government may nationalize the company. While this may not be the end of the world, it is a good way to break the country's line to the world technological circuit. The dynamic advance of countries which are plugged into this world circuit is in marked contrast to the stagnation in many countries professing stale, socialistic doctrines.

However, the world mechanism does need some modification to take into account the frustration or the resentment threshold which is built up by the imported technology. It is only prudent to seek to accommodate to these problems as they arise, and to make the engine of international technological transmission a still more effective instrument of human progress. The difficulty is that when we tinker with a social mechanism, we can never be sure that we are improving it. In this case, when we yield to pressures to do something about technological gaps and brain drain, we may be emasculating some of the vitality of the mechanism itself. Will actions once taken compensate for inequities in the process, or will they subsidize inefficiencies which in reality are delaying adjustments which must take place?

The world technological circuit may be visualized as a series of technological escalators running in parallel with interlocking relationships and feedbacks linking the faster moving with the slower moving. With few exceptions all nations have an escalator, and everyone wants his to move still faster. Everywhere the idea of technology is associated with growth, and though everyone wants technology in principle, there is little understanding of what must be done to get it and great reluctance to make changes to install it.

Broadly speaking, there are two divisions of escalators, the developed and the underdeveloped world. The problems in each set are different, and there are numerous subsets, but whatever modifications may be made in the foreseeable future, basically the same sort of medicine must be prescribed for all. The prescription: When a country is behind in a technology, it must decide to go into the market and buy an existing advanced system, and, in turn, it must make adjustments within itself necessary to utilize that technology. The possibility of becoming a great center of research and development in an area may emerge at a later date as comparative advantage develops. But the pragmatic doctrine of learning to walk before one runs is sound policy for both newcomers and countries which have been superseded in some specialized field.

The primary assumption underlying this line of thought is the conventional wisdom that God helps those who help themselves. In this case the wisdom is correct. Those countries which accept or are prepared to accept an achievement-oriented society are those which will acquire and use effectively the technology. Those who are not willing to accept this orientation or are incapable of making the adjustments will find the new technological world uncomfortable. There is evidence, however, that most countries—leaders and people—accept the achievement-orientation ideal. The proclaimed interest in economic development and growth bears witness to it. Idyllic village development schemes in the tradition of Mahatma Gandhi have been largely rejected. Still, there may be some nations which will opt away from the international escalator achievement pattern. Herman Kahn thinks that China will not follow this path.[2] But for most of the world, the crucial question will not be acceptance of the more efficient escalator idea but how to best effect it, or how to improve the country's technological performance relative to the rest of the world. This means how to transfer the technology efficiently and to learn to cope with the change it entails.

DEVELOPING COUNTRIES

One fashionable statement about technological transfer to underdeveloped countries proclaims that technology is not something to be set down on the dock, uncrated, and put to use. In fact, however, to a considerable degree, it is just that—at least in a generalized sense. As was explained earlier, modern technology consists not only in mere physical objects, but in a complex system which to be truly effective must be obtained and absorbed as holistically as possible. "Uncrating" such a complex begins the local people's experience and necessary adjustment to this system, which must continue through later phases of installation and operation. The technology or the system forces people to make the necessary adjustments to it. The idea of designing technology to fit the people is largely a will-o'-the-wisp.

Naturally, no one is opposed to doing as much as possible to alleviate the friction caused by the introduction of technology. And the people themselves will try to make some adaptations to suit local conditions, but in most cases it is to be expected that these changes will be minimal. When the author asked the Japanese petro-chemical firms what changes were made in the product, after long thought a typical answer would be that "we reduced the size of the containers because we Japanese are smaller than you Americans," or that "we changed the color of the product to a pastel shade." No doubt there are cases where more significant changes took place, but the idea that one must tailor the technology to the conditions of the people and the environment is not, outside of agriculture, a very practical program. Who would develop such tailoring? Research people in the developed countries are too busy with modern technological advances to take time and commit resources to such exotic experimentation.[3] Old technology, thought to be more appropriate because more labor intensive, may well no longer exist, and we delude ourselves that it can be easily resurrected.

In any case the local people do not regard the idea of hand-me-down technology with great favor, and they are right on the economic grounds of competition. In the modern world of standardized products and quality control, old-fashioned, labor intensive methods are often simply not competitive. If one wants to develop an import substitute or an export capability, in, say, retread tires, the way to do it is to obtain the modern machines and methods which make them to standard specification. No one will want the labor intensive tires because even if they could be made, they could hardly match those made by the machines under quality-controlled production. Certainly they could not be exported. The related problem of providing jobs for the population can be met in other ways—small scale industries, handicrafts, and service industries—but it should not be confused with the acquisition of modern, competitive technology. The Japanese precedent always stressed modernization, and the formidable character of Japanese export drives depended on the combination of relatively cheap labor and modern technology. Acquiring technology in the developing countries should stress modern technology and in that way obtain the benefits of leapfrogging, especially in developing competitive exports.

Once the character of the technology is decided, every feasible channel should be utilized as opportunities arise. For some reason, the military channel has been overlooked in the past, yet its great potential has historical precedent, and it is ready at hand in virtually all countries. Military training programs in an underdeveloped country are a substitute for general training, indoctrinating a rural population in the rudiments of hygiene and sanitation, using electrical switches, appliances, and telephones, etc., or, as in one story told to the author, preventing the faucet from being wrenched from the pipe to obtain tap water. In some countries the chief source of such skills as pilot, mechanic, or electronics technician is the armed forces. These personnel are often trained directly or at one remove by the military forces of the United States or of other advanced countries, and when the technicians leave military service, they form the core of the poor country's advance in their area.

There are also important industrial opportunities for those countries cooperating with United States military assistance programs to obtain modern technology direct from the research center itself. Formerly, only countries with considerable industrial base qualified to receive large quantities of supporting technology for repair and overhaul of aircraft and other modern equipment. Today, there is a gathering trend to make the cooperating nations self-sufficient. Thus, Korea has developed an aircraft repair, overhaul, and even rebuilding or manufacture capability comparable to an American facility in California. The military channel is an opportunity to keep abreast of the best modern technology in a vital area with all the implications it carries in the way of technological multiplier effects in other areas.

But one big general opportunity to introduce industrial technology in an underdeveloped country is to induce the international companies which often have heavy military and research and development contracts to locate some part of their business interest in that country. The advantage is that the country gets the technology as a going system of equipment, patents, processes, technicians, task forces of specialists, managerial and professional talent of all kinds as well as the outlets to the world market which the big, international entity has built up. Moreover, the international company has a long-term horizon and plans investment and operations with continuity. Typically nowadays, it is prepared to accept considerable social responsibility for its workers in terms of housing and other amenities. If properly cultivated, the local subsidiary or affiliate can have linkage effects by encouraging related industrial undertakings. Subcontracting to small business may grow up along the lines of the Japanese pattern.

The well-known antagonism to foreign private capital in many countries stemming from colonial heritages and socialist dogmas may be so great that the foreign package of technology is impossible. If so, then the next best thing is some sort of licensing agreement without the much feared "penetration of capital." Again the Japanese pattern provides a precedent. An alert people, using a Japanese-style antenna type of program to send *shisatsudan-* (literally look-and-examine) style teams abroad, can spot appropriate licensing opportunities and induct the technology at home with strategic know-how purchases of assistance. If local entrepreneurship cannot be found, foreign contractors can be hired or some aid obtained on an almost free basis through such Peace Corps-type help as the recently formed group of American retired businessmen willing to lend their experience to benefit developing nations. However, efficient transfers are bought in large projects.

Such strategic purchases of the package system of business, however, must be accompanied with a big push to build inputs in the less developed country's education system—particularly technological education. Debates over the emphasis on general versus technical education are rather painfully academic. If the goal is to acquire technology, technical education must have precedence though a certain amount of basic education is obviously necessary to build on. University education of a liberal arts nature is another luxury which should be limited in favor of turning out engineers.

Science education and training in its more theoretical aspects may be developed, but great care should be exercised in using scarce resources in a scattered approach. No doubt fundamental scientists have a place in any country's structure as wise counselors and models of inspiration for students, but if some care is exercised in deciding on policy concerning what research areas are appropriate to a country, the pay-off is likely to be greater. Other things being equal, a poor country can hardly afford research laboratories in atomic physics, but research institutes in tropical agriculture for a tropical country may be eminently sensible and practical. Agriculture is one area of the economy where the technology does have to be developed to fit the environment. The technical advances of agriculture in developed countries are for the temperate zone. The Paddock brothers' book cites many examples of why simple transfer of technology will not work in agriculture.[4] Therefore, the opportunity for the establishment of research institutes with high pay-off possibilities is great in areas of agriculture. Similarly, where scientific endeavors cannot be duplicated in other areas, investment in research is indicated. Thus, tropical disease might best be studied in a country like the Congo.

Accordingly, the training of scientists should be encouraged in such possible research fields and limited in less appropriate disciplines. Universities which are developed in poor countries should emphasize advance training in such research activities. As to training in foreign universities, students should be sent abroad for advanced study at government expense only when there is a good chance of using them in a capacity in the home country which will permit them to use what they have learned. It is becoming clear that it is often a bad social investment to send students for a complete education overseas in fields selected by the students. They often become dissatisfied with the contrasting situation at home, and they provide recruits for emigration to the developed countries. Restricting and channeling of students into research fields that will pay off for the home country in the case of theoretical science will do much to overcome the wastage of investment which occurs through the brain drain.

The same channeling toward engineering and technical fields must be begun for undergraduates. The primary rule is that they should be educated at home as much as possible. For such education as the students obtain abroad, it seems only basic fair play to work out international political agreements to control the flow of those trained. The spectacle of these students forming a "foreign aid" program for the United States and the advanced countries in which these student contributions may be greater than the aid extended to the home countries approaches the absurd. Further, the reliance on foreigners to fill gaps perpetuates inefficiencies within the advanced countries which should be corrected. The shortage of physicians or graduate students in physics, for example, is only tolerable because of the subsidy of foreign specialists making up the difference. However, the primary responsibility must lie with the developing country to provide job opportunities for the engineers and technologists which it permits to be trained. This also is directly related to getting on with the job of introducing technology in package form at every conceivable opportunity.

That such decisions interfere with the free choice of individuals in the developing countries must be accepted as the cost of modernizing these countries. Science and technology have come down from the past coupled with the tradition of free choice of subject matter. Now, however, it has simply become imperative that social goals be served, and as in all cases where freedom of the individual conflicts with democratically accepted social goals, the latter must have priority. Technology for poor countries is too important to be left entirely to free choice.

Having made this point, it is obvious that whatever system of control may be enacted to channel people from poor countries into more pedestrian, workaday world technologies with immediate payoffs, the human spirit will find ways of evading such controls. There will always be those who will slip through the net and manage to educate themselves to become the purest of mathematicians, and the world may benefit enormously five generations later. This too is desirable. All that is being argued here is that some attempt to channel scarce human resources into more quick-yielding technical occupations is a better policy for poor countries than the present, relatively undirected system which develops overtrained specialists who are often lost in an exodus to the developed world.

This type of policy can best be formulated in the forum of the International Technology Agency which has been suggested in this book. Under its aegis, dialogues between poor and wealthy country representatives entered into wholeheartedly may well develop intelligent and systematic approaches to this kind of problem which will be reasonably satisfactory to all. Restrictions on the intake of students and careful monitoring of their training courses has already been undertaken in programs of foreign student sponsorship under the auspices of such agencies as the Census Bureau in the United States. With such precedents it should not be difficult for the Agency to work out mutually satisfactory curricula and monitoring methods which would effectively transfer specific technologies. When such trained individuals are geared into programs of tailored institution-building in the country under sponsorship of government, universities, business, or the military, the whole process will take on the meaning and vitality which it presently lacks in large measure.

THE DEVELOPED COUNTRIES

In the developed countries, intervention with the freedom of individuals to make their own choices is probably less necessary, but much depends on the willingness of these countries to make the fundamental adjustments necessary to keep people at home. Basically, this means creating a more competitive environment for modern, research-based technology. The potential emigrating engineer is not likely to find the foreign offer so attractive if it is matched, even partially, by opportunities at home.

The heart of the problem is to match the American challenge with a response which puts priority on technology and the research effort behind the technology. Fundamentally, this means two things. First, it means continuing to accept the technological transfer in the kinds of packages from multinational, American-based business which it has been doing in the past; and second, it means creating the conditions for a larger scale research effort in order to match the base which provides a fountain of American corporate innovations. Neither of these courses of action is easy.

The first course has been followed in the past. Since the Marshall Plan period, repeated infusions of American models and techniques have occurred. Europeans have complained of Americanization of their culture again and again, but productivity teams and management consultants opened up new horizons, and sooner or later, the more effective technique or machine was introduced or licensed, and the new, more rationalistic technology prevailed, bringing higher productivity and gains in growth. Moreover, it cannot be overemphasized that this was not a one-way flow, and it did not occur across the board. Thus, the Japanese who imported the best technology took two thirds of its know-how contracts from the United States, but the rest came from Europe. Such international giants as Nestle in Switzerland, Imperial Chemical Industries in England, Philips in the Netherlands, and numerous others developed new products, processes, and techniques presumably superior to those of the United States at the time. Otherwise, the Japanese would not have bought them.

But for those industries in European countries which became Americanized, as the popular expression has it, there were distinct pay-offs in new products, growth, and profits. Growth, as Schumpeter held, proceeds in spurts. After a few years there is a plateau in growth, but licensing, or one's own research and development, brings out more new products, and the growth begins again. New growth springs from new products. Similarly, cost-reducing innovations bring higher profits. Export markets are more easily penetrated with the managerial emphasis on the consumer and with linkage of research with marketing and sales effort. Some of this has been accomplished by changes effected through Japanese-type antennas, or high level "look-and-examine" teams, but in many cases the spur has had to be actual competition from American-owned subsidiaries. In fact, it may be argued that the challenge to overcomfortable ways by American-owned or closely affiliated firms is a boon in disguise to the domestic industry. The Sears, Roebuck and Company store in Barcelona has induced a swarm of Schumpeterian imitators.*

Technological nationalism, however, quickly rears its head. What is remembered are the fierce competition which is engendered, and the successes of the American firms in driving out a few local firms rather than its success in forcing change. There is fear of American predominance in spite of whatever is done. The understanding is widespread abroad that the source of the American effectiveness lies in the massive research and development activity, particularly in the heavy United States government and military contracts which finance a large proportion of this research and development. How can companies compete without similar programs and backing from their governments? The governments of the rich countries taken individually are not so rich as the United States, and they feel that they are not able to mount research and development efforts on a comparable scale. To compete with American firms, supranational conditions for financing large-scale research efforts must be created. Put simply, this means implementation of Prime Minister Harold Wilson's vision of a greater technological Europe; or alternately the development of some large Anglo-American coalition, side by side with the European Common Market, is in order.

Within the framework of the Common Market, there are already precedents of scientific and technological collaboration. These include the European Organization for Nuclear Research, the European Space Research Organization, the European Launcher Development Organization, and the European Coal and Steel Community. These are all rather specialized agencies, as is the British-French collaboration to produce the Concorde, supersonic airplane or the British-German-Dutch enriched uranium project.[5] The European Parliament, the North Atlantic Treaty Organization, and Organization for Economic Cooperation and Development all have committees which have worked on the problems of cooperation, but thus far there has been little progress toward the establishment of large-scale funding of research activities. Yet getting on with the serious effort to become competitive in research and development

* Discussed in the last chapter.

is the long-run way to become more viable and less technologically dependent on the source of American research and development.

With or without such restructuring of research under supranational programs, there is a kind of inevitability in the way international technology transfer will continue to function. It is a well-known principle that when pressed, business interests resort to political devices to accomplish what economic activity in the market place has failed to accomplish. For a time these may have some success, but at the cost of failure to make essential changes. Without protection, temporary equilibriums will be followed by the next wave of investment if it is profitable. Further American encroachment in many industries coupled with an exodus of expensively trained scientists and technicians will continue. Technological nationalism is bound to reassert itself as it has already done in France with uneconomic and duplicative nuclear and space efforts. Bans on free movement of scientists and engineers will be imposed with consequent loss of freedom of choice for an important class of individuals. Yet over the longer run, larger market entities with greater research and development and more competitive technology will come into being in the developed countries as a result of the American challenge. These in turn will challenge American preeminence in markets at home and abroad, as is already happening in many industries.[6] In reverse flow, the market system will transfer technology, reestablish a new equilibrium, and after a time a new cycle of disruption will begin again. On balance, it is a beneficial sequence, and it requires only a little monitoring possibly from an international agency, as proposed in Chapter 8.

THE UNITED STATES

Foreign technology policy for the United States is set in the context of other international policies and problems in its relations with nations developed and underdeveloped. Generally speaking, it pursues a policy of making technology available through business, military, and government channels, and setting the conditions for a good deal of free movement of scientists, engineers, businessmen, and students in and out of the United States. In the case of the developed countries in Europe, it has supported a basic policy of encouraging European economic integration. Closely associated with this idea was the corollary of linking a United Europe with the United States in an Atlantic community of nations. In the case of the developing nations, the policy has been to assist the national planning efforts to generate self-sustaining growth through economic aid, including the encouragement of training their nationals in the United States.

Some success has been achieved in the implementation of this vision, but difficulties have arisen, no small part of which is the failure to recognize the profound, disruptive impact of the new technology elsewhere. Because of its relative absence of traditions and its huge size, the United States is able to absorb the new research-based technology with reasonable ease, but even here the domestic issues of automation, unemployment, and urban decay and unrest, for example, testify to adjustment difficulties in the United States itself. In other countries the institutional difficulties stemming from the infusion of new technology may be even greater.[7] Adjustments take longer and arouse much resentment. Since the United States is at the center of the worldwide generation of new technology, the United States is the target of other nations' resentment of change, and thus it has a scapegoat role to play in the thinking of governments and peoples in many countries.

Nothing the United States can do is likely to change the need for making internal adjustments to the new technology by whatever country to which it is transferred. New, largely military, research-based technology is creating a new kind of world and is disequilibrating the old order to a degree unimagined a few years ago. The proliferation of future forecasts, describing what the world is likely to be twenty-five, fifty, or more years hence, testifies to this demand for information to cope with the emerging new world.[8] Every country with few exceptions is involved in the process of making adjustments, and this is not something that can be eased very much by other countries, no matter how well intentioned. In fact, the openhandedness of the United States in making the technology available so easily is itself a major factor in the difficulties it is experiencing in the flood of competitive imports.

In spite of this uncertain future, there are some parameters in the technological equation that are reasonably constant and predictable. Achievement-oriented men everywhere are going to seek to better themselves, and achievement-oriented managers of corporations will continue to seek opportunities for growth and profit. Socialist and other oppositions will continue to manifest, as they now do, a technological nationalism. Yet in the balance, more rather than less concern for human welfare is likely to emerge.

The United States, perhaps more than most nations, has a keen sense of its welfare being related to the welfare of the world community of nations. It is of course a national entity with only 6 percent of the world's people, and as such it must look after its own interests. But it has committed itself to working out its destiny in such a way that it wants its prosperity and well-being to function within an advancing world of co-prosperity and diffused well-being. Cynics may say that this is the only possible policy for a modern capitalist state among socialist wolves, but the fact is that it is an outgrowth of the profound egalitarian, large-thinking conceptions of traditional America. Henry Ford typified this attitude when he treated his labor not only as a cost, but also as income earners who would buy back his products. The modern multinational company which says that "we want everyone connected with this business to make money—the workers, the management, the suppliers, the agents, the dealers—and at the same time we also want the consumer to receive a better product" is carrying out the tradition. The vitality and greatness of the United States lies in precisely this strength, and it has been chosen to be the spearhead of technological advance and transfer of the new technology in this century.

There is of course always room for improvement, but it is difficult to see how the United States, short of an atomic holocaust or an internal revolution, could or would wish to alter the thrust of technological change, development, and sharing which has been its course. In the short run, some marginal improvements may be in order, but these will be instituted on a fairly experimental, one-step-at-a-time basis. Tampering with the technological engine may well have unforeseen consequences because of the complexities of the social changes going on. Anything that is done should be in line with a policy of measured response and feedback of results. For the longer run, the United States must help to build such international institutions as the International Technology Agency proposed in this book to cope with the problems of international adjustment to technological change.

NOTES TO CHAPTER 11

1. National Bureau of Standards, *Technology and World Trade, Proceedings of a Symposium, November 16–17, 1966* (Washington, D.C.: 1967); University of Denver, Denver Research Institute, *Conference on the Environment and the Action in Technological Transfer,* Snowmass-at-Aspen, Colorado, September 26–28, 1969 (proceedings to be published).

2. "Herman Kahn's Thinkable Future," *Business Week,* March 11, 1967, pp. 115–18. Herman Kahn and Anthony J. Wiener, *The Year 2000* (New York: Macmillan, 1967). Dropouts to the hippie culture also are at least temporary exceptions to achievement orientation.

3. One exception is the research program for adapting technology to the needs of developing countries at Battelle Memorial Institute in Frankfurt, Germany.

4. Paul and William Paddock, *Hungry Nations* (Boston, Mass.: Little, Brown, 1964). The "green revolution" in India and Pakistan depended heavily on dwarf wheats and rice strains, adapted to the local environment after much experiment.—Carroll P. Streeter, *A Partnership to Improve Food Production* (New York: Rockefeller Foundations, 1969), pp. 11–14, 26.

5. *Financial Times,* December 19, 1969.

6. "U.S. Goods Face a Run for the Money," *Business Week,* December 6, 1969, p. 206; "Overseas Investment: Nationalism Crimps the Spending Spree," *ibid.,* p. 202.

7. Many Historic Wonders in Europe Threatened by Modern Hazards," *Wall Street Journal,* December 22, 1969, pp. 1, 17.

8. Robert U. Ayres, *Technological Forecasting and Long-Range Planning* (New York: McGraw-Hill, 1969). One example of more scientific work in this field.

12
QUESTIONS AND PROSPECTS FOR THE FUTURE

This book on technology gaps and transfers inevitably raises more questions than it answers. Yet if the world as we know it is to survive the impact of constantly changing technology, some attempts, however faulty, are in order to provide perspective on what is happening. Following Keynes' dictum in praise of the brave army of heretics who "preferred to see the truth obscurely and imperfectly,"[1] this essay has tried to make a start on visualizing dimensions of an emerging world system of technology. In present day America, where it is becoming fashionable to deny the existence of a technological gap with elusive euphemisms like "innovational disparity," or to de-emphasize the contributions of the military-space complex to the civilian economy, these thoughts may seem to carry the challenge of controversy. Such militancy is not intended. What is offered is, to paraphrase a quotation from Schumpeter,[2] something to start from—an agenda for further research among scholars, and reflection among informed persons everywhere.

DISCERNIBLE NODES

Some of the principal points made in this book are: First, the new, research-based technology emanating largely from military sources is a new phenomenon. It is dimensionally different from the craft technology of former eras. The new technology is of course related to science, but *science* is not a descriptive or broad enough term for the various ramifications of the new technology. Technology is a thing *sui generis*—in a class by itself—which requires this stress. It is better to speak of technology policy than science policy if the relevant emphasis is to be obtained.

Second, military-related research and development is not an effort to be deprecated, but rather to be understood, accepted as a present-day reality, and appreciated. Results of the research transfer into the civilian economy and create new innovations which would not have come into existence without the military-related complex. Today, these new technologies are being conveyed to other countries around the world in what is actually a comprehensive system of technological transfer. The lamentations of "technology gap," "brain drain," and "economic imperialism" are manifestations of the system in operation. Third, the transferred technology is forcing changes in the social and economic structure of countries everywhere. This has been so in the past, and it will continue to be true in the future. The problem is one of coming to terms with the new technology and of better organizing the world as a technological system.

Fourth, each polarity and subset in the world system has problems peculiar to itself. The United States is largely responsible for the kind of technology which is being developed. It is innovational, science-based technology, oriented toward use and application. The United States has grown and flourished under a system of relative freedom of development, and this is true of its government, of business, and of the military. Its actions, however, are causing tremendous repercussions abroad.

For the developed countries, the core of their difficulties is that the elite classes have traditions of pride and prejudices which interfere with adjustments to the new technology. New systems of international cooperation must be devised by these countries in order to use the new technology effectively and beneficially. For the underdeveloped countries, the model is Japan: simple, pragmatic imitation of the best practice in any part of the world. To obtain the new technology quickly and efficiently, they must do as Japanese managers are said to have done: to take as their motto "Find out what the best practice is—and do it better." [3] It is equally appropriate to the developed world, but their pride may obscure understanding and prevent application. Fifth, the polarities of the world technological system are connected by channels. Three of these channels have been discussed briefly in this book: government, military, and business. Each needs to be studied further to discover ways of improvement. The government channel particularly requires the creation of an international technological authority which will study the problems of world technology with detachment. The military channel needs to be greatly improved. Americans must cease being embarrassed by its contribution, or by associating it with wars and weaponry only; instead they need to be more conscious of the value of the military channel, of its peaceful uses, and seek more ways of benefiting from its advantages.

The business channel offers the best opportunity for transferring technology to other peoples and countries as a complete "package." Here we have all the components necessary to move technology from one part of the world to another. Nevertheless, the business channel has many difficulties and problems for its participants. These can, however, be smoothed away by dialogues between representatives of the international companies which have the technology, and the leaders of the underdeveloped world who are wary and somewhat hostile to anything that suggests to them an economic imperialism.

Lastly, though improvements are essential, the present world technological system is a going concern which has the advantages of accomplishment and the disadvantages of creating disturbance. Its accomplishments thus far are nothing short of miraculous. Technologies are developed from the idea-to-creation stage of the laboratory, and are barely in existence for six months before they are found in many parts of the world as practical, commercial activities. However, the very success of the system has evoked the hostility of the established interests and elites in many countries. Adjustment to social and other kinds of change is not easy, but it appears to be the price of technological advance. Which countries will pay the price, and under what time intervals, are questions for the future.

Every society has need of cultural continuity and cannot afford to pay too high a price by disrupting this continuity. Elting Morrison describes the case of the *U.S.S. Wampanoag,* a warship which the United States Navy of a century ago rejected because it was thirty-five years ahead of its time and its introduction too overwhelming for the Navy's human organization.[4] Yet the role of the new technology "in breaking through the crust of tradition and opening up opportunities for individual initiative and creative drives" is crucial for all countries.[5] In the modern world the people of each country must weigh objectively the values of cultural continuity against the values of cultural fluidity, and make their own decisions. For most modern peoples, however, the presumption seems to be in favor of the fluidity side or, phrased alternatively, the new technology forces them to this side. In any case, men of good will everywhere must learn to view the new technological phenomenon with more objectivity and detachment in place of the petty, technological nationalism so prevalent at present.

TECHNOLOGICAL GAPS AND TRANSFER IN PERSPECTIVE

The new world technological system is being formed from the challenges and responses, pressures, tensions, and accommodations which are evolving in its present context. The interactions which have been described have arisen largely spontaneously, and were planned only at a subsystem level, if at all. In a larger sense, what is in existence and what is evolving result from the forces which are grinding inexorably to ends not previsioned or planned.

Military-related, research-based technology is creating changes and problems which are restructuring a new kind of life for mankind. The national cries of technological inferiorities are in some sense harbingers or warnings of things to come. They serve to draw attention to the complexities of the evolving system, and they force us to look ahead ten, twenty, or more years, just as research men are accustomed to do in their respective fields.

It is said that the American Telephone and Telegraph Company visualizes an ultimate communications system which would equip everyone anywhere in the world with a two-way telephone, worn as a wristwatch, capable of quickly contacting any other individual in the world. When a caller cannot reach the other party by this futuristic telephone, he will know the ultimate constraint has intervened—that the person he is trying to communicate with is dead.

Some type of ultimate thinking such as this is needed for a view of the world technological system. Innovations to make man's life and lot simpler, more rational, and more effective are now being developed and transferred across the world. The cries of anguish which are now being heard are due to the birth pangs of the new world technological system. Rather than taking alarm, the world should be appreciative of the "distant early warning system" of more complexities to come. The present book has sought to make a start on outlining some of its dimensions and laying a foundation for continuing dialogues between disparate interests involved in the system. Hopefully, this foundation has raised questions which will germinate interest and stimulate further investigation.

Volumes may be written on the intricacies of technological policy for each of the participant countries. In the United States, few persons except official advisers are thinking in terms of a technological policy for the country. The impetus of a resentful Europe is forcing consideration of technological policy, but the very term is still unfamiliar in the United States. As to Great Britain and other advanced countries, their painful groping toward technological policies has been described briefly in the relevant chapters, but narrow, nationalistic orientations, and the weight of past traditions are only too evident.

Among advanced nations, only Japan has found a reasonable balance between dynamic technological advance and cultural stability. Barely removed from the underdeveloped world, the Japanese maintain both their ancient cultural system and their achievement orientation, and continue to startle the world with their economic growth, much of which is directly related to exports based on technological borrowings. Now Japan is beginning to invest heavily in indigenous research and development.[6] Herman Kahn may well be correct when he predicts that the twenty-first century will be the Japanese century.* Be that as it may, Japan, the avid technological modernizer, in borrowing and utilizing new technology for growth is a model for both the developed and developing countries.

In the latter category the true gap is so great that many believe it is already unbridgeable. In any case in the developing countries, the technological gap and transfer problems are part of the general one of assisting the poor countries to undergo effective technological and economic development in which Americans have tried to assist them in the past. This is turning out to be a very slow process, and the developing countries must be encouraged to do more for themselves and to move more quickly. Perhaps, as now seems likely, national development is linked to individual or group achievement-orientation; if so, more encouragement should be given to helping dynamic people in each country. Furthermore, incentives must be devised to keep highly trained people in productive and creative work in their home countries.

At this writing, some part of the technological transfer problem has been the subordination of technology to a host of ancillary factors. If, as this book has tried to do, technology is given separate identity and accorded priority, it is believed that fresh approaches

---

* See Chapter 7 for citation of Kahn's views.

to closing the poverty gap between rich and poor countries and between the rich and poor people in each country may be possible. There is little point in decrying a harsh technology which forces people to adjust to it. Adjustments have been made to progress in the past, and must be made now and in the future. Understanding of the mechanism of technology transfer and adjustment is an essential preliminary step to improving it.

Every channel of technological transfer now in use must be clarified, updated, and made more fruitful. The military channel particularly deserves the most careful study and attention because it previously has been overlooked even by military men themselves. Hostility toward the military institution must be converted to objective recognition of its positive attributes as a major generator of innovations and a conveyor of international technology.

The question has been raised whether American intellectuals are not overzealous in applying the principle of separatism of military and civilian activities to foreign countries.[7] It is questionable whether this principle is suitable for countries where political traditions of democracy are noted by their absence, and the social structures are inherently authoritarian. Given the priority goal of transferring technology and closing gaps, the trade-off of some political purity for techno-economic advantage may be a flexible, effective course of action. The indigenous military forces in underdeveloped countries are especially suitable for use as agencies to build technological capabilities and transfer technology into their supporting civilian economies from which commercial applications follow naturally. The local military organization is in economic terms a sunk cost in developing countries, and utilizing this investment to transfer technologies to the civilian economy is an eminently sensible agenda for poor countries which devote, often compulsively, large proportions of their national products to defense.

New institutions for promoting technological transfer such as the International Technology Agency proposed in this book must take their place beside the older channels and those proven international institutions in related fields like finance, food, agriculture, and labor. The world needs fresh, objective thinking on the subject of the international technological system, and it will obtain this only if respected, dispassionate agencies staffed with excellent minds are created to increase the benefits of the technological system.

But despite the great need for changed and improved institutions to accommodate technological change, the vitality of the ongoing system as it exists today should not be overlooked or discounted. The system we now have is essentially effective, and whatever its deficiencies, it brings forth new technologies, the vast majority of which are highly beneficial to humanity, often in totally unexpected ways. When permitted to function properly, the present system transfers the new technologies across international borders in a remarkably short time and as effective replicas of the original. The agenda of the future calls for both acceptance and creative response in order to improve the viability of an effective ongoing process.

NOTES TO CHAPTER 12

1. J. M. Keynes, *The General Theory of Employment, Interest, and Money,* (Harbinger edition; New York: Harcourt Brace, 1964), p. 371.

2. Joseph Schumpeter, *Business Cycles,* abridged by Rendigs Fels (New York: McGraw-Hill, 1964), p. viii.

3. See Chapter 3, footnote 15. The success in putting the motto into practice is attested by the fantastic drive of the Toyota Motor to sixth place among world automobile manufacturers. The president of the company says he is aiming at first place, above Volkswagen, Fiat, and the Detroit giants.—*Fortune,* December, 1969, p. 135.

4. Elting Morrison, *Men, Machinery, and Modernization* (Cambridge, Mass.: Massachusetts Institue of Technology Press, 1966), pp. 98–122.

5. Neil W. Chamberlain, "Training and Human Capital," Daniel L. Spencer and Alexander Woroniak, (eds.), *The Transfer of Technology to Developing Countries* (New York: Frederick A. Praeger, 1967), p. 163.

6. Organization for Economic Cooperation and Development (OECD), *Science Policy in Japan* (Paris: OECD, 1967), p. 217.

7. Charles E. Hutchinson, *The Role of the Military in the Transfer of Technology* (Washington, D.C.: U.S. Air Force Office of Scientific Research, 1968), manuscript, p. 15.

INDEX